Elementary Semiconductor Physics

Elementary Semiconductor Physics

H. C. Wright
The Allen Clark Research Centre

VNR
VAN NOSTRAND REINHOLD COMPANY
New York - Cincinnati - Toronto - London - Melbourne

Published by Van Nostrand Reinhold Company Ltd.,
Molly Millars Lane, Wokingham, Berkshire, England

Published in 1979 by Van Nostrand Reinhold Company,
A Division of Litton Educational Publishing Inc.,
135 West 50th Street, New York, N.Y. 10020, U.S.A.

Van Nostrand Reinhold Limited,
1410 Birchmount Road, Scarborough, Ontario, M1P 2E7, Canada

Van Nostrand Reinhold Australia Pty. Limited,
17 Queen Street, Mitcham, Victoria 3132, Australia

Library of Congress Cataloging in Publication Data

Wright, Hubert Charles.
Elementary Semiconductor Physics.

Includes index.
1. Semiconductors I. Title.
QC11.W93 537.6′22 78-26036
ISBN 0-442-30198-7
ISBN 0-442-30250-9 pbk.

Filmset in England by Santype International Ltd., Salisbury, Wilts
Printed and bound by Unwin Brothers Limited
The Gresham Press, Old Woking, Surrey

To Alison and Caroline

Preface

Existing semiconductor texts furnish excellent material at specialist level and thoughtful expositions are available for the intelligent layman, but the serious student, in his early years, appears to have a limited literature at his disposal.

This book is intended to serve those starting a career in semiconductor science or technology, through either an industrial training scheme or a university course. Basic physical mechanisms involved in semiconductor practice are each introduced with a self-contained qualitative description, supported by a simplified quantitative analysis. The effect is then exemplified by its use in a circuit device or in the measurement of a semiconductor property. While the qualitative passages can of course be followed through without mathematical knowledge, the quantitative work requires mathematics up to A' level standard.

I would like to thank Dr. D. Walsh who read, and commented most helpfully on, the first draft of this book; and to acknowledge the stimulus of working with my associates at The Allen Clark Research Centre and the generosity of the management in affording me duplicating facilities. I am very grateful to Mr. B. C. Pope for meticulous proof reading.

H. C. Wright, Blisworth, 1978

Contents

Introduction

In the past 20 years the electronics industry has not only made great advances in the classical realms of telecommunications and domestic entertainment but has generated many completely new technological systems. Much of the great family of computers, sophisticated monitoring and diagnostic instruments for medicine, weapon systems for the armed forces and navigational and landing systems for aircraft has stemmed from one brilliant concept that was put into practice soon after the Second World War.

Basically, the pre-war electronics engineer had only the thermionic valve available for either generating or amplifying electrical power at radio frequencies. If these operations were to be carried out in a solid rather than the vacuum in the glass envelope of the valve, the solid device could be smaller, more robust physically, more efficient, more reliable and faster in its response. Scientists at the Bell Telephone Laboratories in America calculated what properties would be needed in a solid if it were to replace the valve, and found that the group of materials known as semiconductors would satisfy all the electronic requirements. They went on to design, and finally make, the first solid-state electronic valve and christened it the Transistor.

This book deals with the physics that underlies the mechanism of the transistor and other devices that exploit the properties of semiconducting materials.

CHAPTER 1

Electrical Conduction

The most evident characteristic of a semiconductor is the magnitude of its electrical resistivity: this lies between the very low values of metals and the very high ones of insulators. The concept of electrical resistivity, and the factors that control it, are dealt with in this chapter.

1.1 METALS, SEMICONDUCTORS AND INSULATORS

1.1.1 Characterisation by Resistivity

In this chapter we shall divide all materials into three classes according to their ability to conduct electricity. There are two related ways in which this ability may be described. First we consider the electrical resistance between two opposite plane faces of a regular geometrical body made from the material (Fig. 1.1). If these

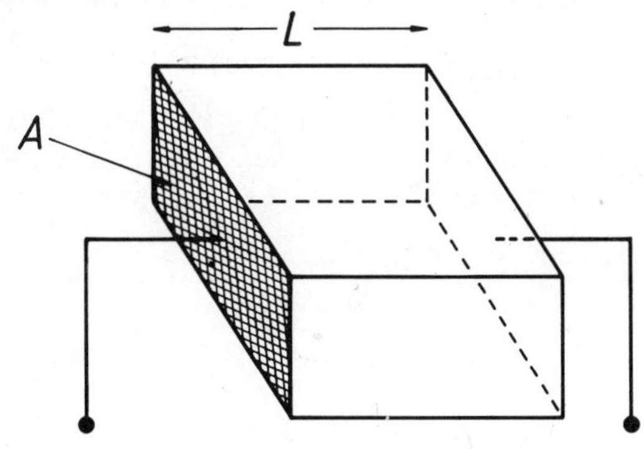

Fig. 1.1 Block of conducting material

plane faces, or electrodes, are separated by a distance L it is clear that the greater L, the greater is the resistance between the electrodes and hence:

$$R \propto L$$

Conversely the electrode area A has an inverse effect on the resistance and this can be seen by supposing it to be doubled, say, when we have in effect two resistors, each of magnitude R, connected in parallel. Thus the resistance is halved and the general rule of proportionality is:

$$R \propto \frac{1}{A}$$

These two proportionalities can be combined to give:

$$R \propto \frac{L}{A}$$

To make this an equation we introduce a constant of proportionality ρ and write:

$$R = \rho \frac{L}{A}$$

or

$$\rho = \frac{RA}{L} \qquad (1.1)$$

Now ρ, unlike R, depends only on the property of the material and is unaffected by the geometry of the particular sample that we are considering. It is in fact the RESISTIVITY of the material. It should be noted that Equation 1.1 shows the dimensions of ρ to be ohms × length and it should be expressed as ohm metre (ohm m). The second way to describe the ability of a material to conduct electricity is by its CONDUCTIVITY. This is merely the reciprocal of the resistivity and is usually written as σ hence:

$$\sigma = \frac{1}{\rho}$$

The dimensions of σ are Siemens (reciprocal ohms) divided by length, conventionally Siemens per metre (S/m).

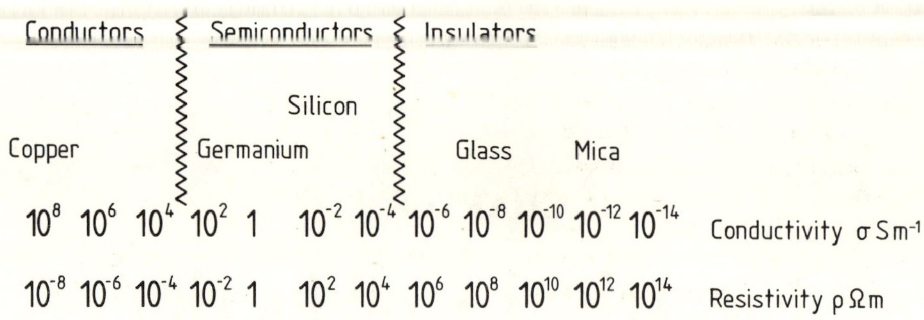

Fig. 1.2 Classification of materials by resistivity or conductivity.

We can now define our three classes of materials:

(1) *Conductors.* Anything with a resistivity of less than 10^{-3} ohm m. This class is made up almost entirely of metals, with copper and silver coming high in the group with their resistivities of 2×10^{-8} ohm m.

(2) *Semiconductors.* Resistivities of between 10^{-3} and 10^5 ohm m will qualify for inclusion in this, the class which will occupy our attention in the bulk of this work.

(3) *Insulators.* Resistivities above 10^5 ohm m allow a material to be classed as an insulator.

This division, shown diagramatically in Fig. 1.2, makes no place for superconductors, the very special group of metals and alloys which display zero resistivity, or infinite conductivity. Further, we shall see that while metals differ fundamentally from semiconductors and insulators the latter two classes can merge into one another. This overlap has led certain linguists to coin the word 'semi-insulator'.

1.1.2 Contribution of Mobility and Free-Carrier Density to the Resistivity of a Material

In order that any electrical conduction can take place at all in a material it must contain electrical charges that are free to move when an electric field is applied. For the moment we will assume that the velocity at which they move is proportional to the magnitude of the applied electric field. Let their velocity be v when a field E is applied then:

$$v \propto E$$

we will introduce the constant of proportionality μ which is known as the MOBILITY to make this an equation:

$$v = \mu E \qquad (1.2)$$

The dimensions of mobility are seen to be velocity divided by electric field. Conventionally the units are $\text{volt}^{-1}\ \text{second}^{-1}\ \text{metre}^2$ (m^2/Vs).

The number as well as the mobility of these charge carriers will affect the current flowing when a field is applied and if there are n such carriers in a unit volume of the material then n is designated the FREE CARRIER DENSITY. There is only one more factor that will affect the resistivity and that is the magnitude of the actual charge carried by each particle. In all the situations that we shall deal with, this charge will be equal in magnitude, although not necessarily in sign, to that of a free electron which is 1.6×10^{-19} coulombs and will be written as q.

Let us now take our original piece of material (Fig. 1.1) and apply a voltage V across it. The field will then be V/L and the velocity of the carriers $\mu V/L$. Thus in unit time a body of charge carriers of length $\mu V/L$ will have moved through each electrode and the number of carriers in this volume will be

$$\text{carrier density} \times \text{length} \times \text{area} = n \times \mu V/L \times A$$

The charge flow through the electrodes is therefore

$$q \times n \times \mu V/L \times A \text{ per unit time}$$

This is by definition the electric current and so we find that when a voltage V is applied a current $qn\mu VA/L$ flows. By definition the resistance R is given by

$$\text{Resistance} = \text{Voltage/Current}$$

$$R = \frac{V}{qn\mu VA/L}$$

but since

$$\rho = RA/L$$

$$\rho = \left(\frac{V}{qn\mu VA/L}\right) \times A/L = \frac{1}{qn\mu}$$

and

$$\sigma = \frac{1}{\rho} = qn\mu \qquad (1.3)$$

This analysis has assumed that the charge carriers are completely mobile or free to move through the material. However, in some materials the charges are loosely tied to some mean position and can only move around this position. Clearly no continuous direct current can flow through such a system but if the applied voltage is regularly reversed the charges can swing to and fro about their central position. This allows an alternating current to flow but as it is a dielectric rather than a resistive property we shall not examine it further.

1.2 DESCRIPTIONS AND USES OF THE THREE CATEGORIES OF MATERIAL

1.2.1 Conductors and Insulators

The good conductors, mostly metals, owe their virtue to the very large numbers of charge carriers which are available. These may be as many as 10^{28} m^{-3} and this great population more than offsets their rather small mobilities which are in the order of 10^{-3} m^2/Vs. The best conductors are silver and copper, their use as wires and connectors in electrical and electronic circuits being familiar. Equally common are the more resistive alloys of nickel and chromium used in heating elements, the tungsten filament in electric lamps and the carbon components used for introducing a precise resistance into a circuit when such is required.

Insulators owe their high resistivity to the small carrier population they have, with most of their charged particles tightly bound to the individual atoms of the material. The very few carriers they have may show quite reasonable mobilities but these will not be large enough to offset their limited numbers. Insulators are of course used to confine the passage of electricity to the desired paths in any circuit.

To sum up then, metals have very large values of n and small μ while insulators have small n and medium μ values.

1.2.2 Semiconductors

In the semiconductor a very wide range of carrier density is found, varying from as little as 10^{19} m^{-3} up to 10^{25} m^{-3}. Mobilities may vary from 10 m^2/Vs to 10^{-4} m^2/Vs. Although no one semiconductor material is sufficiently versatile to show both limits of both parameters the possible resistivities cover many orders of magnitude.

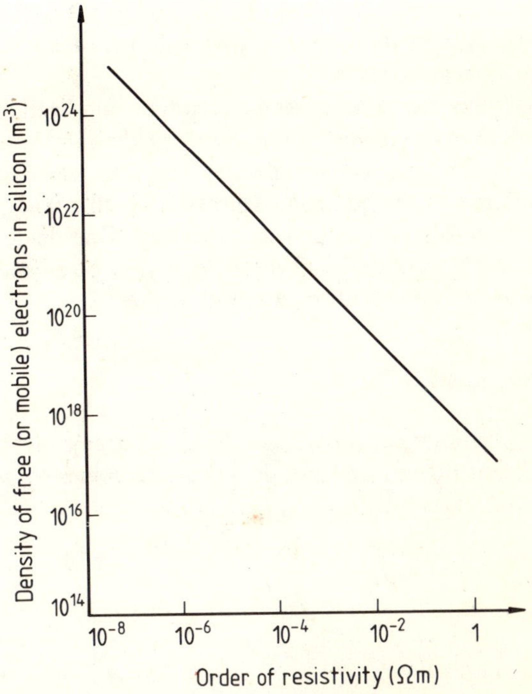

Fig. 1.3 Resistivity of silicon with varying impurity content.

Fig. 1.2 shows the general resistivity grouping and Fig. 1.3 the possible resistivity range for the semiconductor silicon. Semiconductors cannot, however, be made as low in resistivity as a good conductor or as high as a good insulator. It would seem that they have very little interest for, or use to, the electrical or electronic engineer and might as well be set aside as scientific curiosities: a fate that they did in fact endure for quite a period. It will later be seen that their particular value lies in the fact that the number of free carriers can be greatly altered by changing their temperature, directing electromagnetic radiation onto them, or subjecting them to a high electric field. None of these effects takes place very strongly, if at all, in good conductors and insulators. The conductors are already full of free carriers and it is difficult to get any more in, while the removal of a few will make only a small relative change. The insulators on the other hand are reluctant to produce any carriers and those that can be generated are active only for a very short time.

The high mobilities shown by the carriers in most semiconductors are also exploited by devices in which a magnetic field is applied at right angles to the direction of motion of the carrier and deflects its path into a curved form. This is similar to the deflection of the electron stream in a cathode-ray tube. Such effects are also seen in metals but because the mobility, and hence the velocity, of the carrier is low only very very small deflections are obtained.

To sum up then; semiconductors are generally exploited in electronic devices by using the relative ease with which the carrier density can be varied, or modulated, or by the strong deflection which a magnetic field produces in the carrier's trajectory. One device at least uses both of these effects while, it must be admitted, there are some eccentrics that use neither.

PROBLEMS

Express numerical answers with an accuracy of three significant figures, and use the following values for any required constants.

ε_0	the permittivity of free space	8.85×10^{-12} F m^{-1} $= 5.50 \times 10^{7}$ electron charge V^{-1} m^{-1}
h	Planck's constant	6.63×10^{-34} J s
k	Boltzmann's constant	1.38×10^{-23} J °K^{-1} $= 8.26 \times 10^{-5}$ eV °K^{-1}
m	free electron mass	9.10×10^{-31} kg
eV	electron-volt energy	1.60×10^{-19} J
q	electron charge	1.60×10^{-19} C

1.1 The resistance between the ends of a cylinder of semiconducting material is 1500 Ω. If the cylinder is 10^{-2} m long and 2×10^{-3} m in diameter, express the conductivity of the material in S m^{-1}.

1.2 If the material in Question 1.1 has a free-carrier population density of 10^{21} m^{-3} express the mobility of these carriers in m^2 V^{-1} s^{-1}.

CHAPTER 2

Basis of Semiconductor Behaviour

The mediocre resistivity of the semiconductor is not in itself a virtue but it is the trade mark of a substance in which the resistivity changes substantially with temperature, with the application of an electric field or when electromagnetic radiation of a suitable wavelength falls upon it; properties shown only in a small degree or not at all by metals and insulators.

These particular electrical characteristics are a consequence of the manner in which atoms making up a crystal of semiconductor are laid out in regular rows causing them to interact one with another. The behaviour of the electrons in these atoms is dictated by this interaction. In this chapter the individual atoms are described in detail and then the effect of assembling them into their rows, to form the semiconductor crystal, is considered.

2.1 BAND STRUCTURE

2.1.1 The Bohr Atom

In this chapter we shall show that electrons, the charge carriers in a semiconductor, are allowed to have only particular energies, and hence particular velocities. These regions of permitted energies are called bands and our concern will be with the CONDUCTION BAND, which is the highest group of energies a carrier can have without being actually free of the semiconductor crystal, and the band immediately below this which is known as the VALENCE BAND. Between these two bands is the ENERGY GAP which is forbidden to electrons. It is this traffic of the electrons across the energy gap between these two bands and their behaviour in either band that gives the semiconductor its particular characteristics (Fig. 2.1).

We have some way to go before reaching an analytic picture of the band structure, and it will be necessary first to consider some models of a free atom; that is one that does not form part of a crystal but is, say, part of a body of gas.

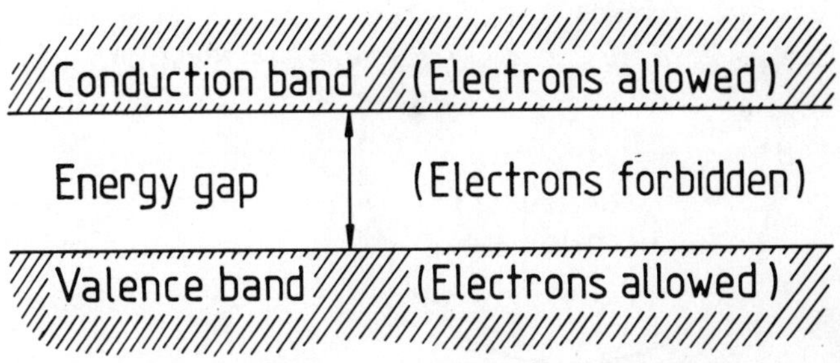

Fig. 2.1 Allowed energy regions for electrons.

Imagine an electron (Fig. 2.2a) and consider one of the lines of force a–x coming from it. If the electron is now moved to a new position a^1 then eventually the line of force will move to a^1–x^1 but the message of the change of position cannot travel more quickly than the velocity of light c. At a time t after the position change any observer farther away from the charge than a distance ct will still think the line of force is at a–x.

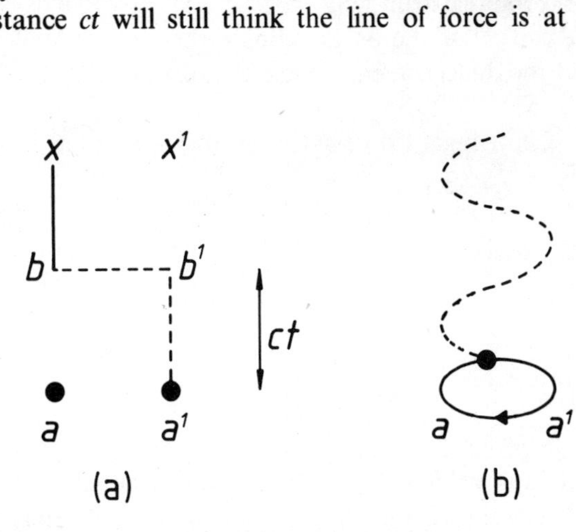

Fig. 2.2 Radiation from moving charges.

The line of force will thus have a kink b–b^1 in it and this kink will travel outwards from the charge with the velocity of light. If the charge now moves back to position a, then a second kink, in the opposite direction, will be generated in the line of force. A regular oscillation between a and a^1 will thus generate a wave train as shown in Fig. 2.2b. If the oscillation takes place at about 10^{14} times a second then the generated wave train will be of visible light.

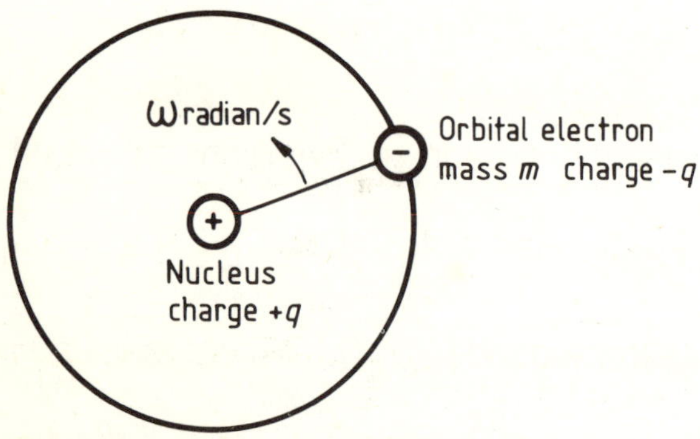

Fig. 2.3 Simple atom model.

One of the early models of the atom proposed a central core or nucleus of positive charge with an electron orbiting around it, the centrifugal force on the orbiting electron being balanced by the coulomb attraction between the positively charged nucleus and the negative charge on the electron. This is analogous to the motion of a planet round its sun where the gravitational force between the bodies replaces the coulomb force between nucleus and electron (Fig. 2.3).

We can relate the orbit radius r, the angular velocity ω and the dielectric constant ε by writing

Centrifugal Force = Coulombic Attraction

$$m\omega^2 r = \frac{q^2}{4\pi\varepsilon r^2} \qquad (2.1)$$

and so

$$\omega^2 r^2 = \frac{q^2}{4\pi\varepsilon m r} \qquad (2.2)$$

Clearly any value of ω can be chosen by adjusting r to satisfy Equation 2.2, or conversely any value of r may be taken by adjusting ω.

2.1.2 The Need for a Quantum Model

This model appears satisfactory until we consider what the effect of the orbiting charge is on some remote observer. It appears to be oscillating from side to side at the rate of $2\pi/\omega$ times a second. Because of this it will be radiating and hence losing energy. No system can afford to lose energy continuously and eventually all atoms, on this model, would collapse.

To overcome this it is necessary to consider, albeit superficially, quantum mechanical ideas. These have their origins in the difficulty that 17th century scientists experienced in reconciling various properties of light. Interference phenomena suggested a wave motion similar to that explaining the propagation of sound, but on the other hand the refusal of light to go round corners was better fitted by a 'projectile' philosophy making light a stream of small particles. Newton proposed the theory of 'Fits', which effectively allowed light to change from one mechanism to the other under certain conditions. This was not, at the time, regarded as one of his better ideas but a few hundred years later it re-emerged in a more precise and mathematically usable form. Today it is held by the QUANTUM MECHANICAL THEORY that all particles have associated with them a wave train or packet which governs their movement and only when they interact with other matter do they behave as particles (Fig. 2.4). One might regard the wave train as a literal train in which the passenger (a particle) is transported. Only on reaching a station (or interacting with matter) does he reappear as a person (or particle). The analogy is useful because it demonstrates another aspect of the theory, which states that the actual position of the particle within the wave train is uncertain and cannot be observed. (The passenger can wander up and down the train and his position within it, when a station is reached, cannot be estimated.) The aspect of this double life which is most evident depends on the energy involved. A massive system such as a speck of dust will be all particle, while an electron will be dominated by its wave properties. This becomes clearer if we write down some formulae.

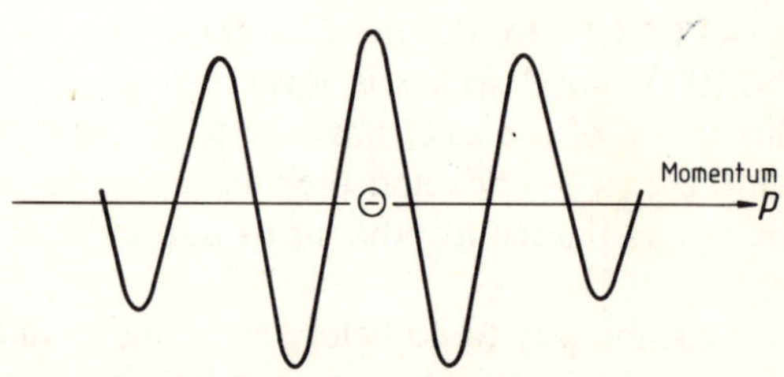

Fig. 2.4 Electron and its wave pocket.

The length λ of the waves within the wave train is inversely proportional to the momentum p of the system,

$$\lambda \propto \frac{1}{p}$$

Putting a constant of proportionality in

$$\lambda = h/p \tag{2.3}$$

The value of h is universal and it is known as PLANCK'S CONSTANT. It has the dimensions of mass × length^2/time.

It will be remembered that we introduced the quantum theory to explain the orbiting electron which did not radiate. It can be seen now that, if the wave train associated with the electron reaches right round the atom, it is not possible to consider it as an orbiting particle anymore. We know it is somewhere in what was previously the orbital path but the quantum theory does not allow us to know where. We must then regard the electron as something stretched around the nucleus like an elastic band.

An obvious consequence of having the wave packet as a continuous band round the nucleus is that there must be a whole number of wavelengths in the band, otherwise there would be a discontinuity, and so the equation

$$2\pi r = n\lambda \tag{2.4}$$

must be obeyed with n being an integer (Fig. 2.5).

The value of λ is known from Equation 2.3, and the momentum p is given by:

$$p = m\omega r \tag{2.5}$$

and so

$$2\pi r = nh/m\omega r \tag{2.6}$$

Equations 2.2 and 2.6 can be written as:

$$\omega^2 r^2 = q^2/\varepsilon m r 4\pi \tag{2.2'}$$

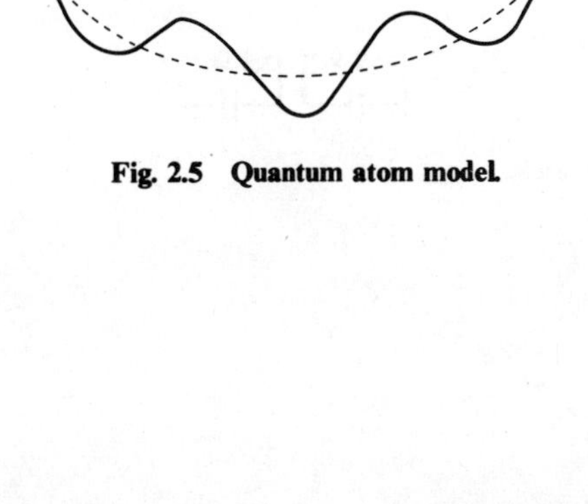

Fig. 2.5 **Quantum atom model.**

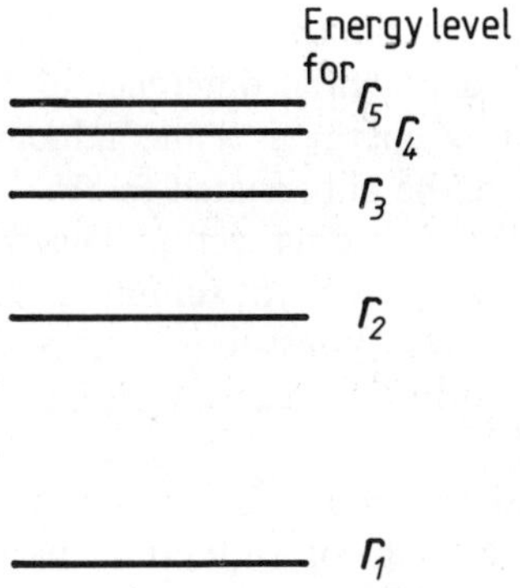

Fig. 2.6 **Energy levels in quantum atom.**

and

$$\omega^2 r^2 = n^2 h^2/4\pi^2 m^2 r^2 \tag{2.6'}$$

Solving these for r gives

$$r = n^2 h^2 \varepsilon/\pi m q^2 \tag{2.7}$$

The family of orbits $r_1, r_2, \ldots$ is obtained by giving n the integral values 1, 2, … (Fig. 2.6).

We will now work out the total energy of the atom.

Total energy = Kinetic energy of electron + Potential energy due to coulomb forces

$$E = \frac{1}{2} m\omega^2 r^2 - (q^2/4\pi\varepsilon r) \tag{2.8}$$

The negative sign in the potential energy term arises from the fact that, in bringing the electron from some remote position to a distance r from the nucleus in order to assemble the atom, the system has done work and hence lost and not gained energy. From Equation 2.2 it is seen that the kinetic energy $\frac{1}{2}m\omega^2 r^2$ is equal to $q^2/8\pi\varepsilon r$ and so Equation 2.8 becomes

$$E = q^2/8\pi\varepsilon r - q^2/\varepsilon r 4\pi$$

Putting in the value for r from Equation 2.7 gives

$$E = mq^4/\varepsilon^2 n^2 h^2 8 \tag{2.9}$$

The difference in energy that exists between different orbits must be examined. Consider the nth and the $(n+1)$th orbits. Let the energies be E_n and E_{n+1} respectively, then

$$E_n - E_{n+1} = \frac{mq^4}{\varepsilon^2 h^2 8}\left(\frac{1}{n^2} - \frac{1}{(n+1)^2}\right) \tag{2.10}$$

For a free hydrogen atom

$$\frac{mq^4}{8\varepsilon^2 h^2} \approx 13 \text{ electron volts}$$

The electron volt is the energy acquired by an electron falling through a potential difference of 1 volt. It is a convenient unit of energy in semiconductor theory and will from now on be abbreviated to eV.

As this and the preceding section have together made a rather long analysis we will summarise them.

(1) A simple atom consisting of a positive nucleus with an orbiting electron should radiate continuously.

(2) To avoid the impossibility of continuous radiation the concept of a quantum particle must be introduced. This assumes the electron to be carried by a wave train, the position of the electron within the train being uncertain.

(3) Because a wave motion and hence a wavelength has been introduced to describe the motion of the electron only certain orbit lengths are permitted. These in turn allow only certain total atomic energies which must satisfy

$$E = -13\left(\frac{1}{n^2} + \frac{1}{(n+1)^2}\right)\text{ eV}$$

where n is an integer.

There are sophistications of the analysis which allow E to be affected by the existence of elliptic orbits, by the precession of the entire orbit about the nucleus and by the rotation of the electron on its own axis. These considerations belong in another book.

Before using the energy levels we have derived to build up the semiconductor band picture we should refer, for the last time, to the suppression of radiation from the orbit electron. Radio engineers may feel that the well established practice of broadcasting by making electrons rush up and down aerials is in danger. This is not so since in the aerial we are concerned only with free electrons and the wave packet which was big enough to reach round an atom is very small indeed compared with the dimensions of an aerial and may be regarded, in this case, as a particle with a precise, although varying, position.

2.1.3 The Tight Binding Approach to Band Structure

If two atoms such as we have just considered are brought close together, their energy levels will interfere one with another and spread into two close levels each differing slightly from the original.

Analogous behaviour is familiar in the case of any pair of resonant systems such as when two tuned circuits are linked together and the original peak response-frequency of each circuit spreads into a broad band for the total system. The more atoms or tuned circuits that are crowded together, the more the original energy level will multiply or the original peak frequency will spread. By the time we have assembled enough atoms to form the regular array of a semiconductor crystal the levels have spread into broad bands separated by gaps which are a consequence of the gaps between allowed orbits in the original free atom (Fig. 2.7).

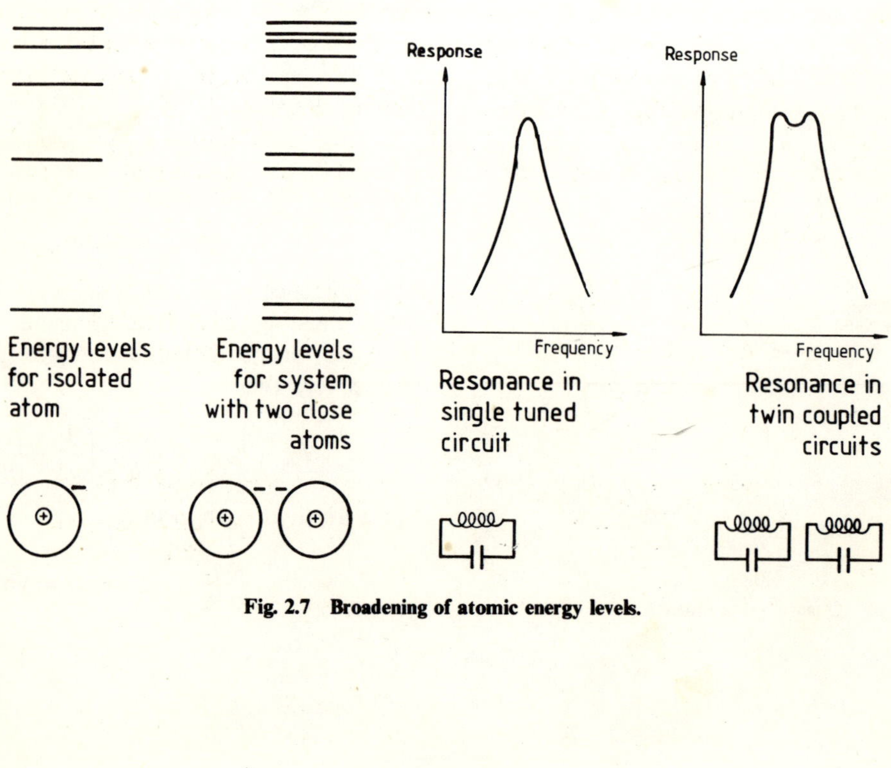

Fig. 2.7 Broadening of atomic energy levels.

Attention has been concentrated on two notional bands because in practice we are concerned only with the highest band and the one lying immediately below it energetically. These are the conduction and valence bands respectively.

The valence band is made up from the orbits which carried the outermost electrons of the free atom and hence controlled its chemical behaviour or valence. The conduction band is so called because electrons in this range of energies have got away from the bonding forces and can be accelerated through the crystal by an electric field and so form the conduction mechanism.

At zero temperature all the available electrons in a semiconductor fall back into the valence band and so conduction is impossible. It has become an insulator. Thus the semiconductor and insulator are closely related. If the material needs a temperature higher than normal room temperature to get many electrons into the conduction band then it will be called an insulator, but if this temperature is high enough to stimulate a fair number into the conduction band it will be designated a semiconductor. Neither will conduct at the absolute zero of temperature (−273 °C).

The manner in which increasing temperature causes more electrons to leap up the energy gap between bands can be understood by considering what is meant by the temperature of a crystal.

We think of the atoms which constitute the material as being in 'fixed' positions forming a regular array or lattice. In fact the atoms will be in constant motion about their mean position at any room temperature above absolute zero. Every so often a vibration will occur which is so vigorous that it shakes an electron off the atom and across the energy gap to the relative freedom of the conduction band. The probability that any particular vibration will be large enough for this to happen is proportional to $e^{-E_g/kT}$ where E_g is the width of the energy gap, i.e. the energy of shaking required, T is the absolute temperature of the crystal and k is a universal constant known as Boltzman's constant. It is seen then that n, the number of electrons in the conduction band, is related to T by

$$n \propto e^{-E_g/kT}$$

We will not yet make the proportionality an equation since the constant needs to be carefully derived.

2.1.4 The Nearly Free Electron Approach

Instead of mentally assembling our crystal lattice from a collection of free atoms we may start with all the atoms in position and see how an electron behaves when it is introduced to this array. Each atom may be simplified to a fixed, positively charged nucleus. Further simplification will be made by ignoring the fact that the lines of atoms really extend in three directions and considering only a single line which will form a primitive one-dimensional crystal (Fig. 2.8). If an electron is now 'pushed' along this line by an electric field it will experience a rather complicated, changing set of forces. When it is at a it will be retarded by nucleus 1; as it approaches b it will be accelerated by attraction to nucleus 2. We will solve what would be a very complex problem, especially in three dimensions, by simply ignoring the detailed behaviour of the electron and saying that, on average, it is accelerated through the crystal as if it had some mass which is other than that which it displays when moving in free space. This is called the EFFECTIVE MASS and is usually smaller than the free mass. For example, an electron in a crystal of germanium shows a value of 9×10^{-32} kg compared with the 9×10^{-31} kg it would have shown in free space. By giving the electron this fictitious effective mass we are able to treat it as a particle obeying Newtonian mechanics.

It will be recalled that in the 'Tight Binding' approach various bands of allowed energies for an electron

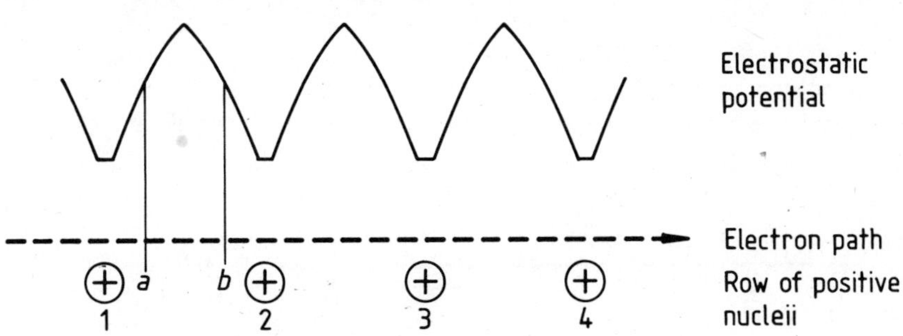

Fig. 2.8 Potential distribution in one-dimensional crystal.

appeared as a simple consequence of their permitted orbits in the constituent of the crystal. It is not so evident from a qualitative treatment of the Nearly Free Electron picture that such bands will appear, until we remember that the electron has a wave disturbance associated with it (Equation 2.3); this wave packet has to negotiate the periodic array of atoms in the crystal lattice.

As the electron passes an atom it will stimulate that atom to radiate with the frequency of the wave packet. At certain frequencies the combined radiation from the regularly spaced atoms will add up to a zero value; that is, destructive interference will take place. The atom array may be compared with an optical grating which prevents the propagation of certain light wavelengths in any given direction.

Because of this, certain electron wavelengths, and hence energies, will be blocked off, giving the now familiar gaps in the permitted energies.

2.1.5 The Density of States

In this section we shall determine how many different energy values an electron may take up in a given range of energy from E to $E + dE$ when the electron is confined in a crystal semiconductor. To do this it is necessary to use two physical principles known respectively as the PRINCIPLE OF INDETERMINANCY and the PAULI EXCLUSION PRINCIPLE. In Section 2.1.2 we saw the rather vague life the electron led wandering up and down in its wave packet. We cannot observe its velocity v or its position x precisely. If the uncertainty in v is Δv and in x is Δx, then the principle of indeterminancy states that

$$m\,\Delta v\,\Delta x \geqslant h \qquad (2.10)$$

We shall not justify this but merely point out that the more precisely the position is known, i.e. if Δx is very small, the greater the uncertainty in the velocity, i.e. Δv is large, and of course vice versa (Fig. 2.9).

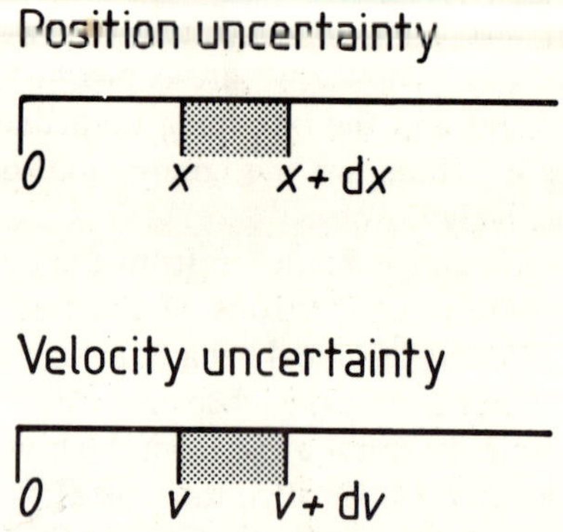

Fig. 2.9 The principle of indeterminacy.

The Pauli Exclusion Principle can be more easily justified. It states in effect that no more than two electrons in an atom can have exactly the same energy. This follows logically when we recall the electron as being wrapped round its atom in a certain orbit and, at first, we would think that only one electron could exist in that orbit. However, because it is assumed that the electron can spin in two ways on its axis, each orbit accumulates two electrons, one with each spin direction. The principle is extended from electrons in an atom to electrons in a crystal, which can be regarded loosely as a large, complex atom.

We are now in a position to make a calculation of allowed energies in the crystal.

Suppose an electron is constrained to move only along a line of length ΔL (Fig. 2.10). The uncertainty is its position is ΔL and so, from the inequality 2.10, Δv, the uncertainty in its velocity, is given by

$$m\,\Delta v\,\Delta L \geqslant h$$

Consider only velocities lying between v_1 and v_2, say. The number of separately distinguishable velocities will be (Fig. 2.10)

$$\frac{v_2 - v_1}{\Delta v} = \frac{m\,\Delta L}{h}(v_2 - v_1) \qquad (2.11)$$

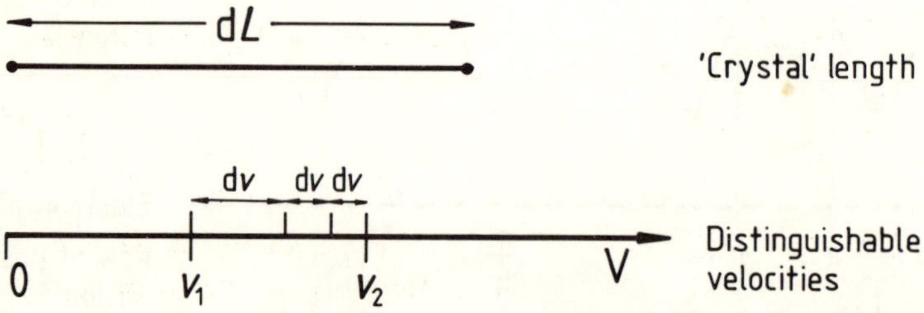

Fig. 2.10 Distinguishable velocities in one-dimensional crystal.

If we now allow the electron to move in a three-dimensional cube of side ΔL instead of merely along a line, all possible velocities must lie in a spherical shell of inner radius v_1, and outer radius v_2 (Fig. 2.11). The volume of the shell in 'velocity space' is

$$\tfrac{4}{3}\pi(v_2{}^3 - v_1{}^3)$$

Now each velocity must be separated from its nearest neighbour by a distance $h/m\,\Delta L$ by Equation 2.10. It therefore requires a volume in 'velocity space' of $h/(m\,\Delta L)^3$. The number of all owed velocities between v_1 and v_2 is therefore dN' where

$$dN' = \frac{\text{Volume of shell}}{\text{Volume required for each velocity}}$$

$$= \frac{4}{3}\pi(v_2{}^3 - v_1{}^3)\Big/\left(\frac{h}{m\,\Delta L}\right)^3$$

$$= \frac{4}{3}\pi\left(\frac{m\,\Delta L}{h}\right)^3 (v_2{}^3 - v_1{}^3)^3 \qquad (2.12)$$

If a cube of unit side is considered $\Delta L = 1$ and, remembering that each velocity may be shared by two electrons, the number of electrons per unit volume becomes dN where

$$dN = \frac{8\pi}{3}\left(\frac{m}{h}\right)^3 (v_2{}^3 - v_1{}^3) \qquad (2.13)$$

Writing the energy associated with each velocity as

$$E_1 = \tfrac{1}{2}mv_1{}^2$$

$$E_2 = \tfrac{1}{2}mv_2{}^2$$

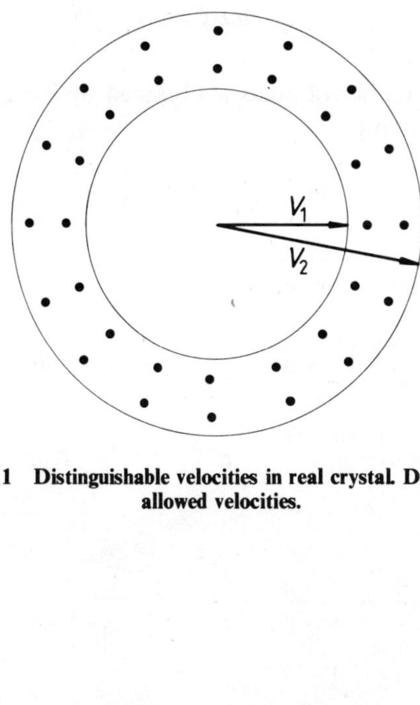

Fig. 2.11 Distinguishable velocities in real crystal. Dots denote allowed velocities.

gives

$$dN = \frac{16\pi\sqrt{2}}{3}\left(\frac{\sqrt{m}}{h}\right)^3 (E_2^{3/2} - E_1^{3/2}) \qquad (2.14)$$

If we allow E_1 to approach E_2 so that

$$E_1 = E$$

$$E_2 = E + dE$$

then Equation 2.14 becomes

$$dN = \frac{m}{h^3}\sqrt{(2mE)}8\pi\, dE \qquad (2.15)$$

This value of dN is then the number of electrons, having energies lying between E and $E + dE$, which can be accommodated in unit volume of a semiconductor crystal.

2.2 CARRIER POPULATION IN THE BANDS

2.2.1 The Fermi–Dirac Statistics

If we take this unit volume of crystal at the absolute zero of temperature and place n electrons in it they will have to take up energies according to Equation 2.15, filling up the available energy 'slots' from zero up to some value which we will call E_F (Fig. 2.12). The value of E_F can be calculated by writing

$$\int_0^{E_F} dN = n$$

but

$$\int_0^{E_F} dN = \int_0^{E_F} \frac{m}{h^3}\sqrt{(2mE)}8\pi\, dE \qquad (2.16)$$

E_F is called the FERMI ENERGY and at zero temperature all energy states below it are filled with electrons and all those above it are empty. If the temperature is now raised to a finite value T then according to the FERMI–DIRAC STATISTICS each of the energy slots reduces its capacity for holding electrons by the fraction

$$1/(1 + e^{(E - E_F)/kT})$$

and equation 2.16 becomes:

$$n = \int_0^{E'} \frac{\frac{m}{h^3}\sqrt{(2mE)}8\pi\, dE}{1 + e^{(E - E_F)/kT}}$$

E' is the energy at the top of the conduction band but this may be taken as ∞ as it will be very large compared with kT and the value of the integral from E' to ∞ will be small.

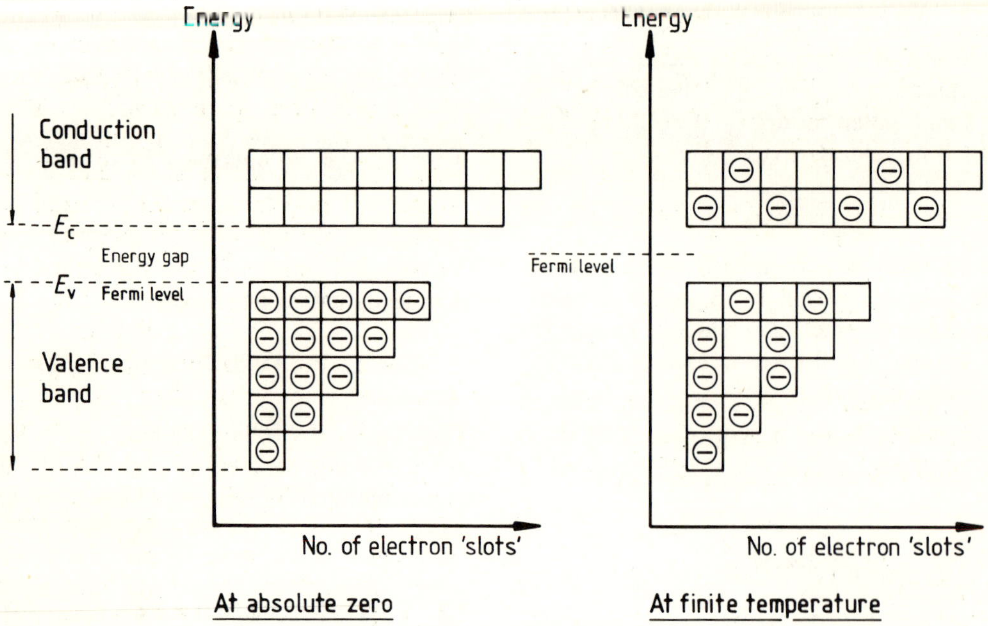

Fig. 2.12 Available electron sites in crystal.

As $(E - E_F)/kT$ is much greater than unity in most semiconductor problems we may write

$$1/(1 + e^{(E-E_F)/kT}) \approx e^{-(E-E_F)/kT}$$

and so n becomes

$$n = \int_0^{\infty} \frac{m}{h^3} \sqrt{(2mE)} 8\pi \, e^{-(E-E_F)/kT} \, dE \qquad (2.17)$$

2.2.2 The Positive Hole

When the temperature was raised, some of the electrons that had previously filled the valence band completely were forced by the reduced accommodation in each energy slot to move up right across the energy gap into the previously empty conduction band. Now the holes that are left behind will apparently move in the wrong direction when an electric field is applied. We say apparently because in fact it is the motion of the remaining electrons in the normal direction that causes the holes to move in the opposite way. This is similar to the upward movement against gravity that the bubble in a spirit level makes when the fluid flows down. For the same reason, mathematical convenience, we will regard the holes as behaving like carriers of positive charge in a semiconductor rather than work out the motion of the many electrons which are still left in the valence band. So for every electron squeezed up into the conduction band, a positive hole having its own effective mass is left behind to conduct electricity in the crystal. The effective mass of this hole will differ from that of the electron because of the different conduction mechanism involved.

Having accepted the hole as a real entity we can work out the position of the Fermi level for a semiconductor at some finite temperature T by twice using Equation 2.17, to obtain the number of electrons and then the number of holes, remembering these must be equal as every electron in the conduction band has left a hole behind it in the valence band.

Integrating Equation 2.17 over the conduction band gives

$$n = \int_{E_c}^{\infty} \frac{m}{h^3} \sqrt{(2mE)} 8\pi \, e^{-(E-E_F)/kT} \, dE$$

and p the number of holes is obtained by integrating a similar equation

$$p = \int_0^{E_v} \frac{m}{h^3} \sqrt{(2mE)} 8\pi \, e^{-(E-E_F)/kT} \, dE$$

On integration these give

$$n_1 = 2\left(\frac{2\pi m_n kT}{h^2}\right)^{3/2} e^{-(E_c-E_F)/kT} = N_c e^{-(E_c-E_F)/kT} \qquad (2.18a)$$

$$p_1 = n_1 = 2\left(\frac{2\pi m_p kT}{h^2}\right)^{3/2} e^{-(E_F-E_v)/kT} = N_v e^{-(E_F-E_v)/kT} \qquad (2.18b)$$

where N_c and N_v are written for the conduction and valence band density of states respectively. We now emphasise the fact that we are implicitly considering pure or INTRINSIC material by placing the subscript 1 on values of n and p. The possibility that the effective mass of the electron could differ from that of the hole is accommodated by writing m_n and m_p respectively. By equating (2.18a) and (2.18b) we obtain

$$E_F = \frac{1}{2}(E_c + E_v) - kT\frac{3}{4}\log_e \frac{m_n}{m_p} \qquad (2.19)$$

Unless m_n and m_p are greatly different, it is seen that the Fermi level falls nearly in the centre of the forbidden energy region. Multiplying Equations 2.18a and 2.18b together gives

$$n_1{}^2 = n_1 p_1 = 4\left(\frac{2\pi\sqrt{(m_n m_p)}kT}{h^2}\right)^3 e^{-(E_c - E_v)/kT} \quad (2.20)$$

Notice that the Fermi level does not appear in this expression so that the product of the number of holes and electrons in any given semiconductor depends only on the temperature. An exception may occur if the material is very impure when it may become DEGENERATE or metallic in its behaviour.

2.3 THE BEHAVIOUR OF IMPURITIES

2.3.1 Donor Impurities

So far, with the exception of this warning about degeneracy, we have been considering pure materials. However, most devices require the use of a semiconductor which contains a very precisely controlled amount of impurity. Supposc one of the atoms in the lattice of the material is replaced by a foreign atom which has more electrons in it than the original atom (Fig. 2.13). These 'spare' electrons will not be used for bonding the atom into position, as would all the electrons in the native atom, and they will be easily stimulated into the conduction band by the thermal vibrations of the crystal. In fact just in the neighbourhood of the impurity atom the crystal behaves as if its forbidden energy gap had shrunk from $E_c - E_v$ to a new value $E_C - E_D$. The spare electron cannot be used for conduction until it has been excited into the conduction band because, unless there are very many of the impurities, they will be isolated in the crystal.

It is possible to gain an approximate value for $E_C - E_D$ by regarding the impurity as a hydrogen-like atom with one effective electron (the 'spare' one) immersed in an electron 'gas'. Now we saw that the energy of an electron in a free hydrogen atom was about 13 eV; thus it would need this amount of energy to free the electron. However, Equation 2.9 shows that this energy depends on the value of m/ε^2. If the system were placed in the semiconductor we would need to replace m by m_n and multiply E by ε_r, the relative dielectric constant of the semiconductor material. Hence we would expect that

$$(E_C - E_D) \approx 13 \times \frac{m_n}{m\varepsilon_r^2} \text{ eV}$$

If ε_r is of the order of 10 and m_n, say $0.1 \times m$, then $(E_C - E_D) \approx 0.01$ eV.

If the temperature is so low that very few electrons can be excited from the valence into the conduction band and so the only effective electron source is the impurity atoms in the material, i.e. if $(E_C - E_D) \ll (E_c - E_v)$, then we may assess the number of electrons

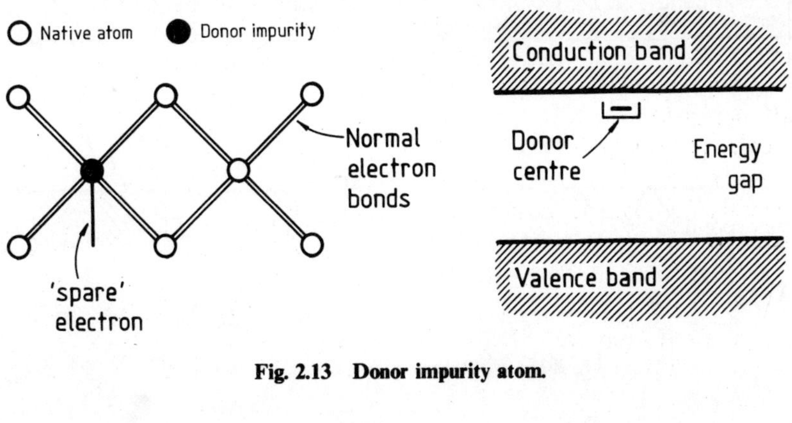

Fig. 2.13 Donor impurity atom.

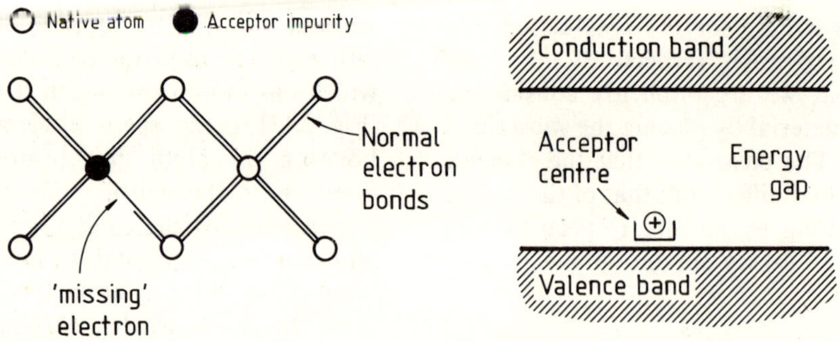

Fig. 2.14 Acceptor impurity atom.

available. Let there be N_D donor atoms per unit volume then by analogy with Equation 2.20

(Number of electrons)^2
= (Density of states product) $e^{-(E_C-E_D)/kT}$

$$n^2 = N_C N_D e^{-(E_C-E_D)/kT}$$

or

$$n = \sqrt{(N_C N_D)}\, e^{-(E_C-E_D)/2kT} \quad (2.21)$$

2.3.2 Acceptor Impurities

If an impurity having too few electrons in its atom for the necessary bonding instead of too many is introduced, it will attempt to make up the deficiency by taking an electron from a native atom near to it (Fig. 2.14). This will leave a hole which behaves exactly as does the hole left when an electron is excited into the conduction band. Such an impurity is known as an ACCEPTOR. By repeating analyses similar to these in the last section we obtain the approximate value of $E_A - E_v$, the activation energy of the impurity, and p, the number of holes to which a density N_n of acceptor atoms will give rise, as

$$(E_A - E_v) \approx 13 \times \frac{m_p}{m}\frac{1}{\varepsilon_r^2}$$

$$p = \sqrt{(N_v N_A)}\, e^{-E_A/2kT}$$

2.3.3 Material Containing Both Donors and Acceptors

If the semiconductor contains both donor and acceptor centres then at low temperatures we may assume that electrons from the donors will fall into the acceptors until either all the donors are empty and N_A acceptors have been filled, if $N_D > N_A$, or that all the donors are filled and N_D acceptors have been emptied if $N_A > N_D$ (Fig. 2.15).

The two possible situations are then:

(1) If $N_D > N_A$ the material behaves as if there were $N_D - N_A$ donors centres present and

$$n \approx \sqrt{[N_c(N_D - N_A)]}\, e^{-E_D/2kT}$$

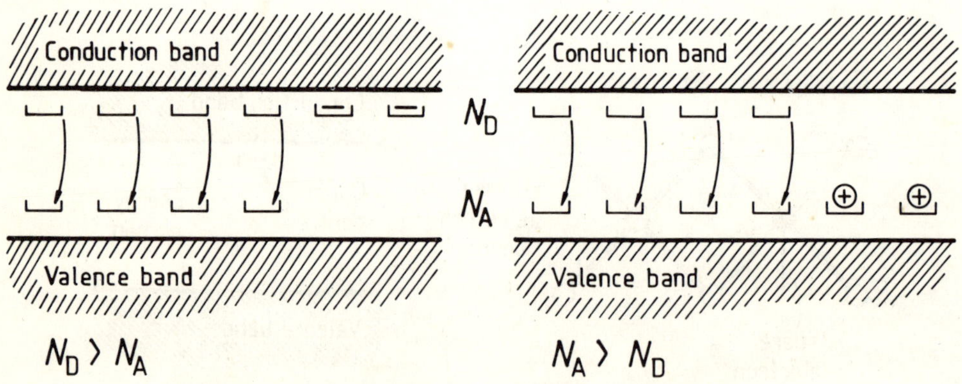

Fig. 2.15 Crystal containing donors and acceptors.

(2) If $N_A > N_D$ the material behaves as if $N_A - N_D$ acceptors are present and

$$p \approx \sqrt{[N_v(N_A - N_D)]}e^{-E_A/2kT}$$

The exact values of n and p can be obtained from the expressions:

$$n - \frac{N_c N_v e^{-E_g/kT}}{n} = \frac{N_D}{1 + [2\,e^{E_D/kT}]\frac{n}{N_c}} - \frac{N_A}{1 + [2\,e^{-E_A/kT}]\frac{N_c}{n}} \quad (2.22)$$

$$p - \frac{N_c N_v e^{-E_g/kT}}{p} = \frac{N_A}{1 + [2\,e^{(E_A - E_g)/kT}]\frac{N_v}{p}} - \frac{N_D}{1 + [2\,e^{(E_g - E_D)/kT}]\frac{p}{N_v}} \quad (2.23)$$

where E_g is written for $(E_c - E_v)$.

These take full account of the fact that the conduction band is constantly exchanging electrons with the donor levels, the acceptor levels and the valence band, and similarly the valence band is exchanging holes with acceptors, donors and conduction band. The full equations 2.22 and 2.33 simplify to the value of n_1 and p_1 for a pure material by writing $N_A = N_D = 0$ and to the approximate expressions for n and p obtained above when $(E_D - E_A) \gg kT$ this implying that the levels are too far apart for carriers to be excited between them by thermal energy.

2.4 EXAMPLES OF SEMICONDUCTING MATERIALS

2.4.1 The Elemental Semiconductors

The elemental semiconductors are shown in their places in the periodic table in Table 2.1. Because most devices require material with an energy gap of about 1 eV at normal room temperatures the most intensively used elements are silicon (with an energy gap of 1.2 eV), germanium (0.76 eV) and selenium (1.5 eV).

Carbon in the form of diamond has been the subject of considerable academic investigation but its 5.2 eV energy gap gives it little commercial use. The remaining semiconducting elements, tellurium, boron, black phosphorous, grey tin, arsenic and antimony, are mostly used in compounds with other elements.

Table 2.1
Position of semiconducting elements in the periodic table.

		Group		
IIIB	IVB	VB	VIB	VIIB
Boron*	Carbon*	Nitrogen	Oxygen	Fluorine
Aluminium	Silicon*	Phosphorus*	Sulphur*	Chlorine
Gallium	Germanium*	Arsenic*	Selenium*	Bromine
Indium	Tin*	Antimony*	Tellurium*	Iodine*
Thallium	Lead	Bismiuth	Polonium	Astatinel

* Semiconducting elements.

2.4.2 Compound Semiconductors

Although industry's biggest use is for material with an energy gap in the region of 1 eV we shall see in Chapter 5 that a group of devices used for detecting infrared radiation requires energy gaps from 0.5 down to 0.1 eV or even less. Where none of the elements shows the precise gap required we are forced to compound pairs of elements, neither of which needs to be a semiconductor. As examples we may consider indium antimonide (InSb) which has a band gap of 0.2 eV and lead telluride (PbTe) with 0.3 eV. In many cases one of the two components of the compound may itself be replaced by a compound or alloy. The lead in PbTe may, for example, be replaced by a lead-tin alloy to make the three-component, or ternary, compound lead-tin telluride (Pb|SnTe). The energy gap of this compound

Table 2.2 Properties of Some Common Semiconductors

Material	Symbol	Band gap at room temperature (eV)	Mobility at room temperature (m²/Vs) Electron	Hole
Silicon	Si	1.04	0.2	0.05
Germanium	Ge	0.63	0.4	0.2
Selenium	Se	1.5	$\sim 10^{-4}$?	$\sim 10^{-4}$?
Tellurium	Te	0.34	0.3	0.7
Gallium arsenide	GaAs	1.4	0.8	0.03
Indium arsenide	InAs	0.33	3.3	0.02
Lead telluride	PbTe	0.29	0.03	0.03
Indium antimonide	InSb	0.16	8.0	0.13
Cadmium sulphide	CdS	2.4	0.02	0.002
Zinc oxide	ZnO	3.2	0.1	

varies from 0.3 eV with pure lead to practically zero when a 60–40 mixture of lead-tin is used.

An improvement in mobility may also be shown by the compound material over the elemental semiconductors. The electron mobility of gallium arsenide (GaAs) for example is $\sim 0.8\ m^2/Vs$ compared with $\sim 0.3\ m^2/Vs$ for silicon, while InSb shows the impressive value of $100\ m^2/Vs$ when used at a temperature of 77 °K.

The many combinations of elements that have been made into semiconducting materials would make a tedious list and so a number of the more common examples, together with band gaps and mobilities where these have been established, are displayed in Table 2.2.

PROBLEMS

For values of required constants, see Problems to Chapter 1.

2.1 An electron is travelling in free space with a velocity of $10^6\ m\ s^{-1}$. Calculate the wavelength of the associated wave train.

2.2 What is the energy required to excite the electron in a free hydrogen atom into its second orbit? Express the answer in eV.

2.3 An electron passes between two electrodes 5×10^{-3} m apart. How closely can the magnitude of its velocity be estimated?

2.4 The effective mass ratio of a free electron in a semiconducting material is 0.1. How many such electrons, having an energy of less than 0.1 eV can be found in a cubic metre of the material at the absolute zero of temperature?

2.5 A semiconductor with a forbidden energy gap of 1 eV has an intrinsic carrier concentration of $9 \times 10^{16}\ m^{-3}$ at a room temperature of 300 °K. To what value will the concentration fall if the semiconductor is cooled to 200 °K?

CHAPTER 3

Carrier Mobility and Lifetime

It has been shown that some of the electrons can escape from their parent atoms for a short time and travel freely about the crystal. If an electric field is applied, the free electrons all tend to drift together in the direction of that field and this movement of charged particles constitutes an electric current. In this chapter the speed of their drift and the time for which they remain free are discussed.

3.1 CARRIER MOBILITY

3.1.1 Analytic Derivation of Mobility

At normal temperatures the semiconductor crystal is a scene of violent activity. Electrons are in ceaseless motion by virtue of their 'thermal' or heat energy. Their paths follow a random zig-zag line as they collide with the crystal atoms, which are themselves swinging ceaselessly around their lattice positions. In addition to these collisions abrupt deflections of the electron path occur due to passage near charged impurity atoms (Fig. 3.1).

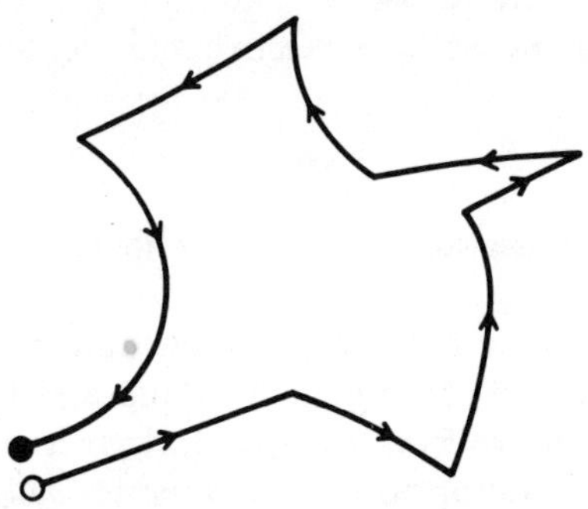

Fig. 3.1 Electron path in crystal: collisions suggested by angular changes of direction, deflections suggested by curvilinear changes of direction.

When an electric field is applied the entire collection of randomly moving carriers drifts along in the field direction.

In order to assess the mobility we must estimate the velocity of this collective drift. To do this we will examine the average time that elapses between collisions.

If the average time between collisions is τ then we must expect some intercollision times to be greater and some to be less than τ. The relation between the number dn in a population of n carriers which have an intercollision time between t and $t + dt$, and the average intercollision time τ is

$$\frac{dn}{n} = \frac{1}{\tau} e^{-t/\tau} dt \tag{3.1}$$

This relationship, shown in Fig. 3.2., is general for random processes and we will prove it but merely note two of its properties:

(1) The sum of all fractions for all possible values of t must be unity:

$$\int_{t=0}^{t=\infty} \frac{dn}{n} = 1$$

(2) The average of all possible values of t should be τ:

$$\frac{\int_{t=0}^{t=\infty} t\, dn}{n} = \tau$$

The reader may wish to check these two integrations.

We will now evaluate the distance travelled, under the acceleration of a field E, for that group of carriers that have a collision time between t and $t + dt$.

$$\text{Distance} = \tfrac{1}{2}\,\text{acceleration} \times t^2$$

$$= \frac{1}{2}\left(\frac{qE}{m_n}\right)t^2 \tag{3.2}$$

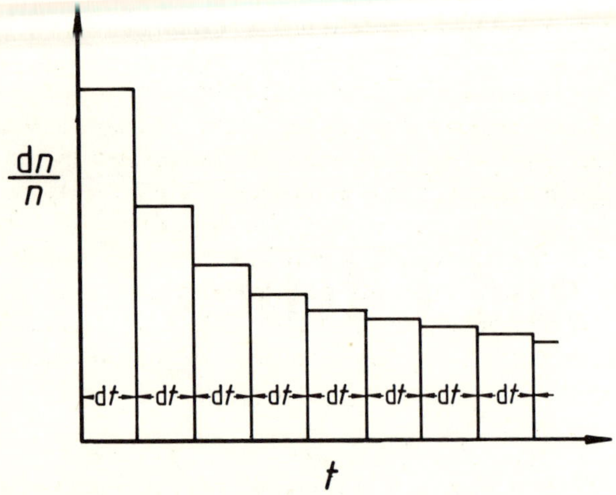

Fig. 3.2 Fraction of electron population having given intercollision time.

The number of carriers which have this collision time t is

$$dn = n \times \frac{1}{\tau} e^{-t/\tau} dt$$

Therefore the total distance dS travelled by the group is

$$dS = \frac{1}{2}\left(\frac{qE}{m_n}\right)t^2 \times n \times \frac{1}{\tau} e^{-t/\tau} dt$$

The total distance S travelled by the entire population is obtained by summing dS for all values of t

$$S = \int_{t=0}^{t=\infty} dS = n\left(\frac{qE}{m_n}\right)\tau^2$$

The total time T for all the collisions to take place is given by summing the product of all the carriers multiplied by their collision times

$$T = \int_{t=0}^{t=\infty} t \, dn = n\tau$$

The average drift velocity for the entire carrier population is S/T and so

$$\text{Average drift velocity} = S/T = \frac{qt}{m_n}\tau$$

The mobility μ is defined as the average drift velocity when $E = 1$ and so

$$\mu = \frac{q\tau}{m_n}$$

This will be our basic equation for estimating the mobility and we will generalise it by introducing both the collision and deflection processes already mentioned.

If a collision with a lattice atom occurs every τ_L seconds and a deflection by an impurity atom every τ_I seconds then in one second there will be $(1/\tau_L) + (1/\tau_I)$ collisions and deflections. This means that the average time between events of either sort will be

$$\frac{1}{\left(\frac{1}{\tau_L} + \frac{1}{\tau_I}\right)}$$

Writing this average time as τ_A gives

$$\tau_A = \frac{1}{\left(\frac{1}{\tau_L} + \frac{1}{\tau_I}\right)} = \frac{\tau_L \tau_I}{\tau_L + \tau_I} \tag{3.4}$$

By substituting this value of τ_A for τ in Equation 3.3 we obtain the total mobility μ_T :

$$\mu_T = \frac{q}{m_n}\left(\frac{\tau_L \tau_I}{\tau_L + \tau_I}\right) \tag{3.5}$$

If there were no lattice collision then $\tau_L \to \infty$ and

$$\mu \to \frac{q\tau_I}{m_n}$$

This is defined as the impurity mobility μ_I hence

$$\mu_I = \frac{q\tau_I}{m_n} \tag{3.6}$$

By assuming that there are no lattice collisions and so letting $\tau_L \to \infty$ we obtain the lattice mobility μ_L

$$\mu_L = \frac{q\tau_L}{m_n} \tag{3.7}$$

Substituting from Equations 3.6 and 3.7 in Equation 3.5 gives

$$\frac{1}{\mu_T} = \frac{1}{\mu_L} + \frac{1}{\mu_I} \tag{3.8}$$

A similar analysis can be made for the hole mobility by substituting the hole effective mass m_p for m_n.

3.1.2 The Variation of Mobility with Temperature

If the temperature is increased the motion of the atoms in the crystal will become more rapid and the distance which they move from their central positions becomes greater. In addition to this the thermal velocity of the electrons will increase. These two changes will have opposite effects on τ_I and τ_L. The frequency of lattice collisions will clearly increase since the atoms are oscillating more rapidly and sweeping out a greater volume while, in addition, the electrons themselves are travelling faster. By virtue of this τ_L decreases with temperature and it can be shown that

$$\tau_L = \frac{A}{m_n^{3/2} T^{3/2}} \tag{3.9}$$

where the constant A depends on the properties of the material. Conversely the value of τ_I increases with temperature since the greater the thermal energy of the carrier the less it is affected by the deflecting coulombic force of the impurity atom.

A sophisticated analysis gives for τ_I the value:

$$\tau_I = \frac{8\sqrt{2}}{\pi\sqrt{\pi}} \cdot \frac{\varepsilon^2 (kT)^{3/2}}{Nq^3 m_n^{1/2}} \Big/ \ln\left(1 + \frac{3\varepsilon kT}{q^2 N^{1/3}}\right) \tag{3.10}$$

Where the material has an impurity density N and ε is the absolute permittivity of the material.

We will now use Equations 3.6–3.10 to write the total mobility in full:

$$\frac{1}{\mu_T} = \frac{1}{qA} m_n^{5/2} T^{3/2} + \left[\frac{\pi\sqrt{\pi} N q^2 \log_e\left(1 + \dfrac{3\varepsilon kT}{q^2 N^{1/3}}\right)}{8\sqrt{2}\varepsilon^2 k^{3/2}} \right] m_n^{1/2} T^{-3/2} \tag{3.11}$$

At low temperatures μ_T is dominated by the impurity component and increases with increasing temperature, until, as the temperature rises, lattice collisions become dominant and the mobility will begin to fall again (Fig. 3.3).

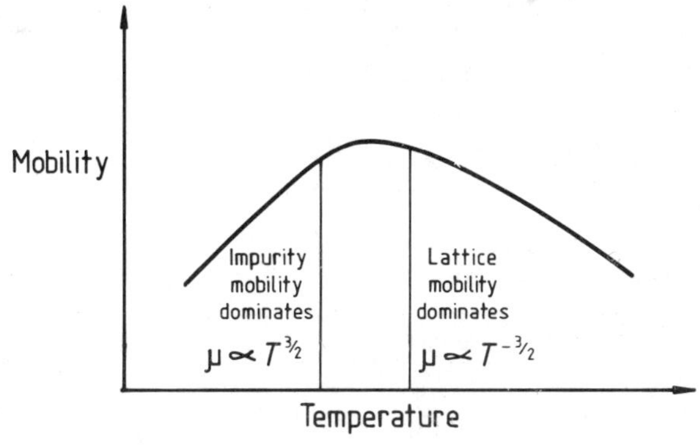

Fig. 3.3 Carrier mobility as a function of temperature.

3.1.3 Effect of Electrical Field Strength

In the last section we ignored the possibility that the increased velocity of the carriers due to the applied field could alter the collision time τ. It was assumed that the thermal velocity greatly exceeded the drift velocity, but this will not be true for very high electric fields.

This thermal velocity V_t may be obtained by considering the kinetic energy of a molecule in a body of gas. Classical theory gives this as $\frac{3}{2}kT$. We may regard the electron as forming part of an electron 'gas' and having a similar energy, hence V_t must satisfy

$$\tfrac{1}{2} m_n V_t^2 = \tfrac{3}{2} kT \qquad \text{or} \qquad V_t = \sqrt{(3kT/m_n)}$$

Suppose we now increase the field until the drift velocity approaches the thermal velocity, i.e.

$$\mu E \approx \sqrt{(3kT/m_n)} \tag{3.12}$$

then τ, the collision time, will begin to decrease. If λ is the average distance between collisions or the 'mean free path' then

Mean free path = Average drift velocity × Collision time

$$\lambda = E \frac{q}{m_n} \tau^2$$

$$\tau = \sqrt{(\lambda m_n / Eq)}$$

$$\therefore \quad \mu = \frac{q\tau}{m_n} = \sqrt{(q\lambda / m_n E)} \tag{3.13}$$

Thus, from Equation 3.12 it is seen that at a field approximately equal to

$$\frac{1}{\mu} \sqrt{(3kT/m_n)}$$

a 'mobility' obtains which, by Equation 3.13, is proportional to $E^{1/2}$. 'Mobility' has now been granted quotation marks because our original concept of it as a constant proportionality between applied field and drift velocity no longer holds.

At even higher fields a second limiting mechanism sets in, causing a complete saturation of velocity. This occurs when the energy of the field-accelerated carrier is large enough to set the semiconductor crystal into vibration with its collisions. When this occurs the energy the carrier derives from the electric field goes into

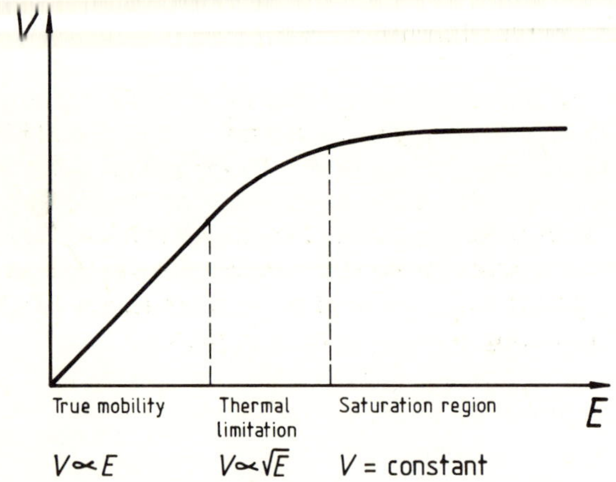

Fig. 3.4 Carrier velocity as a function of electric field.

vibrating the crystal lattice rather than increasing the drift velocity. As the temperature of the crystal is gauged by the violence of its atomic vibrations this operation is equivalent to heating the semiconductor. This special energy that is needed to set the crystal into vibration was first defined when analyses of the specific heat of crystalline solids were being made by Debye. Because of this history the energy is usually expressed in terms of a temperature, the Debye temperature θ_D

$$\text{Critical energy} = k\theta_D$$

The critical electric field for the velocity saturation is thus approximated by

$$\text{Carrier energy} = k\theta_D$$

$$\tfrac{1}{2} m_n(\mu E_c)^2 = k\theta_D$$

$$E_c = \frac{1}{\mu}\sqrt{(2k\theta_D/m_n)} \qquad (3.14)$$

The three regions of drift velocity as a function of applied field are shown in Fig. 3.4.

3.2 CARRIER LIFETIME

3.2.1 The Concept of Carrier Lifetime

The population of any large country is kept at a stable level by virtue of constant and equal processes of birth and death. Similar factors operate to keep the carrier population in a semiconductor at a fixed value. The thermal vibrations of the constituent atoms will regularly excite an electron up into the conduction band leaving behind a free hole. This 'birth rate' is balanced by existing free electrons falling back into the valence band to combine with free holes to form fixed nonconducting electrons (Fig. 3.5). If the generation rate is g electrons and holes per cubic metre per second and the lifetime of a hole-electron pair is τ (not to be confused with the collision time of the last section) then a stable population density of n free carriers per cubic metre will be obtained where

$$g = \frac{n}{\tau} \qquad (3.15)$$

If g is now increased to $g + dg$ by, for example, shining appropriate radiation on the material and so supplementing the thermal vibrations with radiant energy, then n will increase to $n + dn$ where

$$g + dg = \frac{n + dn}{\tau}$$

Taken with Equation 3.15, this gives

$$dn = \tau\, dg \qquad (3.16)$$

The increase dn in the carrier population when the specimen is irradiated is important as the basis of the phenomenon of photoconductivity, discussed in Chapter 5.

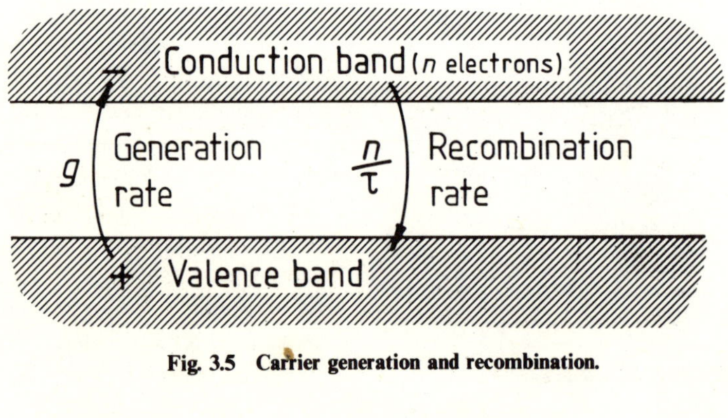

Fig. 3.5 Carrier generation and recombination.

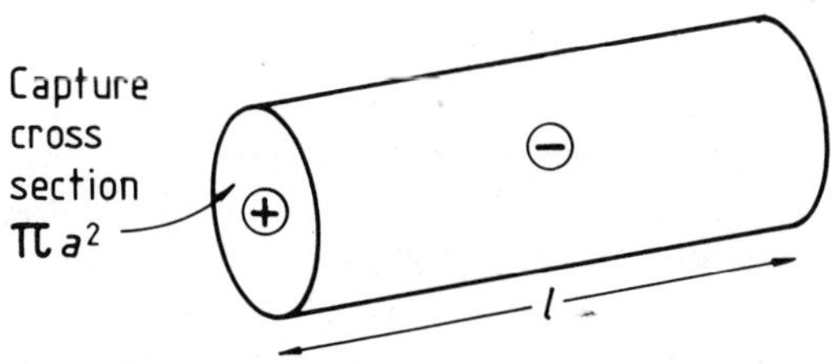

Fig. 3.6 Capture volume swept by carrier.

Although τ cannot be directly computed we can relate it to a concept known as the CAPTURE CROSS-SECTION of the carrier. Suppose that any hole-electron pair that approach to within a distance a of one another recombine. Then we may regard the electron as a point object and the hole as having an area πa^2 for recombination. This area is the capture cross section. If a hole with its cross-sectional area πa^2 travels a distance l it will on average enclose one electron in the volume it has swept out if

$$\pi a^2 l = \frac{1}{n} \tag{3.17}$$

If the thermal velocity of the hole is v then it will travel a distance l in a time l/v (Fig. 3.6). From Equation 3.17 though, one recombination will take place if

$$l = \frac{1}{n\pi a^2}$$

Hence l/v the time for one recombination is

$$\tau = \frac{l}{v} = \frac{1}{v}\frac{1}{n\pi a^2} \tag{3.18}$$

We now write Equation 3.15 as

$$g = \frac{n}{\tau} = n^2 v\pi a^2 \tag{3.19}$$

The form of this equation with its square law dependence on n arises because the number of free holes equals the number of free electrons, a consequence of the material being assumed pure or intrinsic.

3.2.2 Traps and Recombinations Centres

We have seen that impurities in the material can act as sources of free electrons (donors), or, by their ability to capture electrons, as sources of holes (acceptors). The donors were assumed to be energetically close to the conduction band and the acceptors near the valence band but this situation could be reversed. If an acceptor-type centre were to be situated close to the conduction band then it would capture electrons from that band rather than from the valence band. Such a centre then acts as a TRAP for free electrons and analogously a donor centre close to the valence band traps free holes (Fig. 3.7).

The presence of these traps will decrease the lifetime of the appropriate carrier since, once trapped, the carrier will be unable to contribute to the conductivity of the material. Suppose there is a density N_t of such traps each with a capture cross-section A, then the lifetime τ_t due to the traps will be given a modification of Equation 3.18.

$$\tau_t = \frac{1}{vN_t A}$$

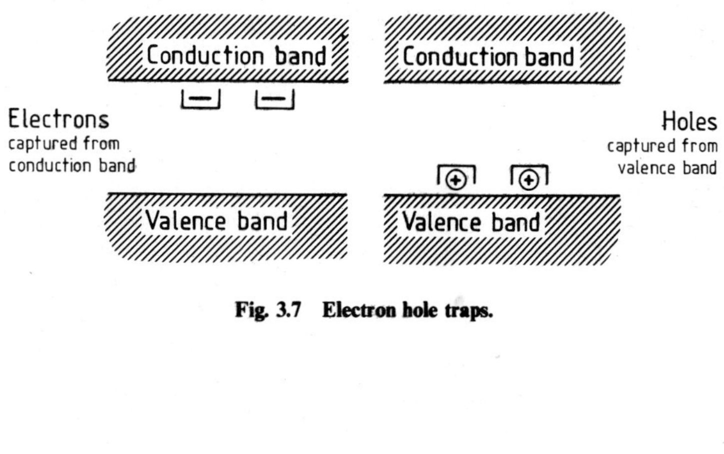

Fig. 3.7 Electron hole traps.

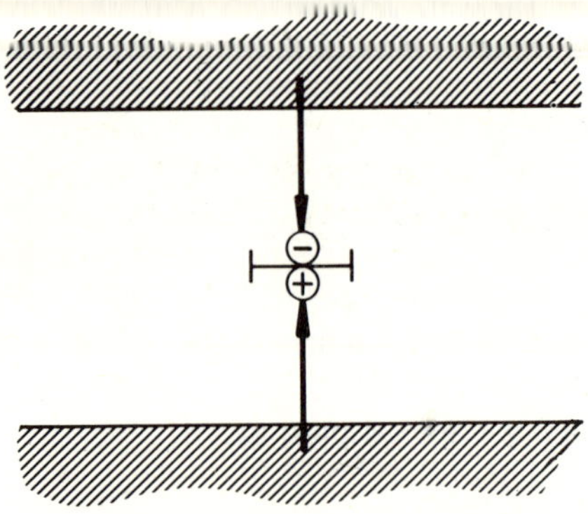

Fig. 3.8 Recombination centre.

If these traps are near the centre of the forbidden energy gap then, after capturing one type of carrier, say an electron, the centre would become negatively charged and have a high probability of subsequently capturing a hole, i.e. losing its captured electron to the valence band. The net effect of the two captures will be the removal of a hole-electron pair from their respective bands. It forms in fact a quick route for the pair recombination discussed in Section 3.3.1 (Fig. 3.8). These RECOMBINATION CENTRES must also reduce the lifetime of the carriers and so τ, the lifetime of a carrier, depends on the density of traps, recombination centres and carriers of the opposite type.

Before closing this chapter we can attempt a crude estimate of the capture radius of a trap once it has made its first capture and become ionized. If an electron approaches a charged centre so closely that its kinetic

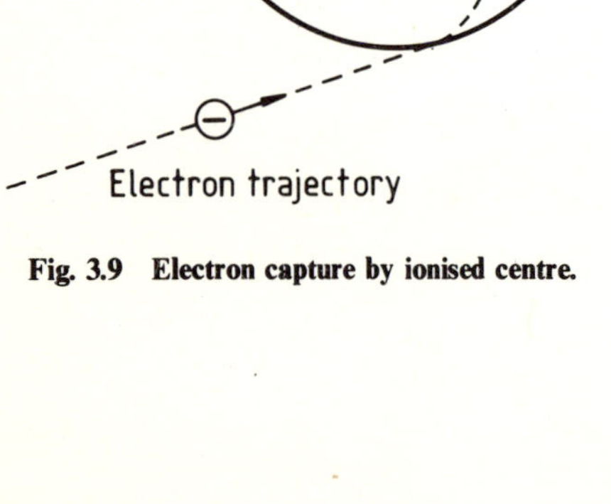

Fig. 3.9 Electron capture by ionised centre.

energy is less than the energy due to the coulomb attraction between electron and centre then it will be drawn into the centre and captured (Fig. 3.9). The critical distance a which just allows the electron to pass without capture is given by

Kinetic or thermal energy of electron = Energy due to Coulomb attraction

$$\frac{3}{2}kT = \frac{q^2}{a\varepsilon 4\pi}$$

$$a = \frac{q^2}{6\pi\varepsilon kT} \qquad (3.20)$$

This is the capture radius of the trap and if $\varepsilon \approx 12\varepsilon_0$ and $T = 300$ °K it will be approximately 5×10^{-7} m. ε_0 is the permittivity of free space.

PROBLEMS

For values of required constants, see Problems to Chapter 1.

3.1 If the average intercollision time for an electron is 10^{-3} s, what fraction of the population will have an intercollision time greater than 2×10^{-3} s?

3.2 A pure semiconductor has a lattice intercollision time of 10^{-3} s. When impurities are added to the material the mobility falls to half its original value. What is the interimpurity collision time?

3.3 A pure semiconductor has a mobility of 0.1 m^2 V^{-1} s^{-1} at 300 °K. To what temperature must it be cooled to increase the mobility by 50%?

3.4 At 300 °K the drift velocity of an electron remains proportional to the applied electric field until this is increased to a value of 10^6 V m^{-1}, after which it becomes proportional to the square root of the applied field. If the temperature is increased to 400 °K, to what value may the field be increased before the mobility again ceases to be constant?

3.5 What is the maximum drift velocity attainable by an electron in a semiconductor having a Debye temperature of 280 °K if the effective mass of the electron is 9×10^{-32} kg, and the mobility is 0.1 m^2 V^{-1} s^{-1}?

3.6 An intrinsic semiconductor has a hole population of

10^{15} m^{-3} at 300 °K. If the effective mass of a free electron in the material is 9×10^{-32} kg and the generation rate for hole-electron pairs is 10^{25} m^{-3} s^{-1}, what is the capture cross-section for hole-electron recombination?

3.7 What is the lifetime of a free hole or electron in the material described in Question 3.6?

3.8 In Section 3.2.1 the capture radius of an ionised trap is calculated to be 5×10^{-9} m. Use this value to compute the permittivity of free space.

CHAPTER 4

Junctions and Junction Devices

If two different semiconducting materials are joined together, a very strong electric field is built up across their junction. The behaviour of such a junction, when an electric current is passed through it, is the basis of most common semiconductor devices.

The transistor, which may be regarded as the founding father of today's electronic systems, consists basically of two of these junctions placed very close to one another. Current passing through one junction leaks, or diffuses, into the second junction and this interaction allows the transistor to amplify.

This chapter describes the fundamental physics that gives rise to the behaviour of the single junction and the transistor as circuit elements.

4.1 p–n JUNCTIONS

4.1.1 Introductory

The best known and probably the most successful application of semiconductors lies in the transistor. This device, invented in the late 1940's by a group of scientists at the Bell Telephone Laboratories, has gradually supplanted the thermionic valve in most applications. To understand its operation we must consider the simple p–n junction.

Suppose, in an imaginary experiment, that a bar of semiconducting material having a large density of donor impurities to make it rich in free electrons (n type) is brought into intimate contact with a similar piece of material rich in holes (p type). The thermal motion of the electrons in the n-type material will carry them into the p-type bar but as there are no free electrons (or only very few) in this region there will be no counter flow of electrons back into the n-type material. This DIFFUSIVE FLOW will denude the n-type side of the junction of the electrons making it positively charged and build up a negative charge in the p region. These two charged regions give rise to an electric field which causes an electron drift back across the junction (Fig. 4.1). Similar considerations can be applied to the drift and diffusion of holes in and out of the p-type material. Equilibrium is reached when an electric field has been built up which causes drift currents exactly balancing the diffusive flows. If now an external voltage is applied across the junction in such a way as to decrease this electric field the diffusive flow exceeds the drift current and, as the voltage is increased, very large diffusion currents will begin to flow. The free carriers which provide these currents come from the bulk of the two semiconductor regions, n and p type, remote from the junction.

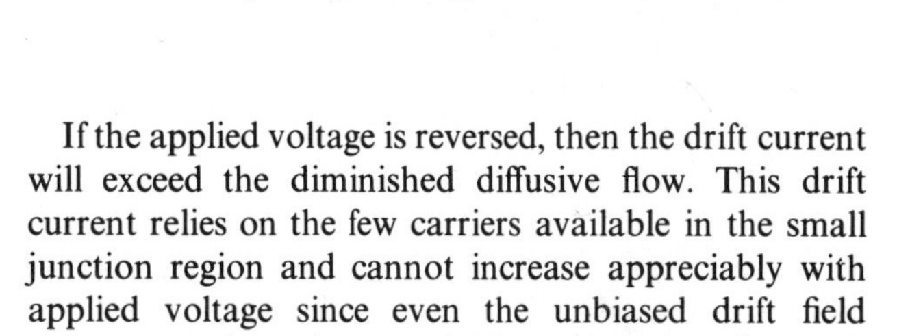

Fig. 4.1 p- and n-type materials in contact.

If the applied voltage is reversed, then the drift current will exceed the diminished diffusive flow. This drift current relies on the few carriers available in the small junction region and cannot increase appreciably with applied voltage since even the unbiased drift field removes carrier pairs as quickly as they are generated. The net current is known as the reverse current and it is qualititatively evident that it will be much smaller than the diffusion or forward current (Fig. 4.2).

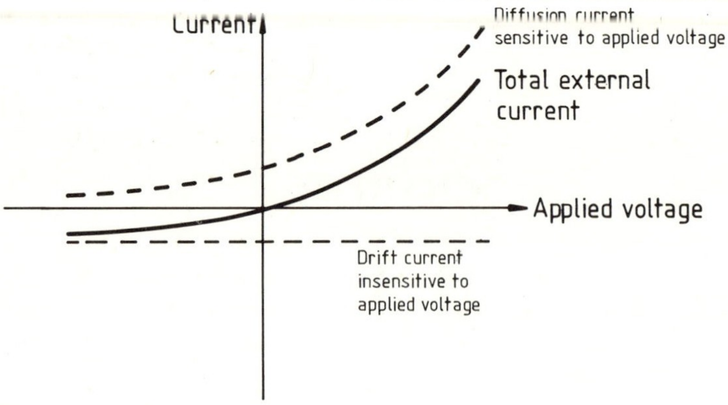

Fig. 4.2 Diffusion and drift currents in junction.

When we imagined the junction at its moment of inception it was seen that most of the carriers diffused away from the junction region leaving a DEPLETION REGION which may be crudely regarded as a slice of insulating material as there are so few current carriers within it. Such an insulating layer, with the conducting areas of the undisturbed material on either side, forms a rather leaky capacitor. This junction capacitance, together with the asymmetrical current-voltage relationship just discussed are the significant characteristics of the junction or diode.

To make a qualitative analysis of junction behaviour we shall need to use three relationships:

(1) If the density n of mobile carriers in a semiconducting bar varies along the length x of the bar, then carriers will flow from regions of high density to those of low density. The rate $\mathrm{d}N/\mathrm{d}t$ at which carriers will flow past a plane in the bar is proportional to the density gradient $\mathrm{d}n/\mathrm{d}x$ of carriers at that plane

$$\frac{\mathrm{d}N}{\mathrm{d}t} \propto \frac{\mathrm{d}n}{\mathrm{d}x}$$

The coefficient of proportionality D which we introduce to make this an equation is known as the diffusion coefficient

$$\frac{\mathrm{d}N}{\mathrm{d}t} = -D\frac{\mathrm{d}n}{\mathrm{d}x}$$

The negative sign allows for the fact that carriers flow in the direction of decreasing density. Variations of n in directions other than x are ignored (Fig. 4.3).

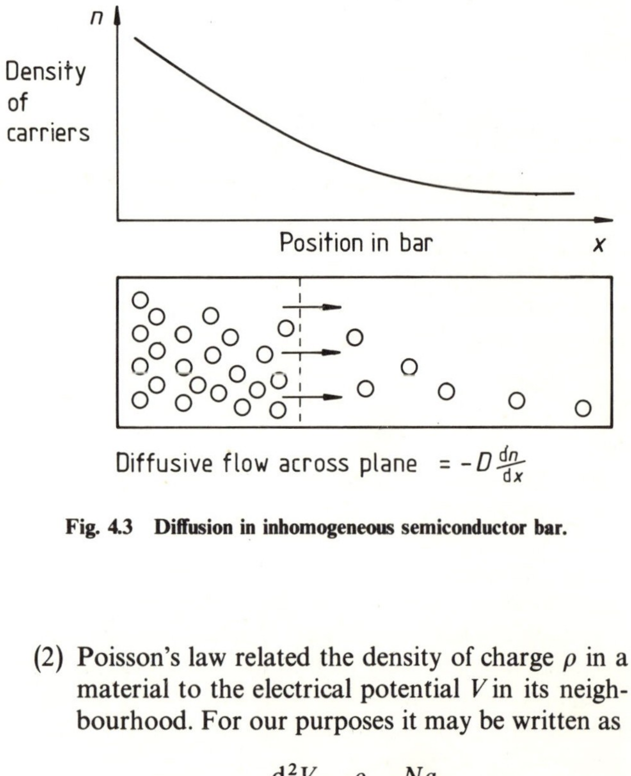

Fig. 4.3 Diffusion in inhomogeneous semiconductor bar.

(2) Poisson's law related the density of charge ρ in a material to the electrical potential V in its neighbourhood. For our purposes it may be written as

$$\frac{\mathrm{d}^2V}{\mathrm{d}x^2} = \frac{\rho}{\varepsilon} = \frac{Nq}{\varepsilon} \tag{4.2}$$

Here ε is the absolute permittivity of the charge-carrying material and variations in V in directions other than x are ignored.

(3) If a potential step of magnitude V is introduced into the semiconductor, the ratio of the carrier densities on either side of the step is related to V by the equation

$$n_1/n_2 = \mathrm{e}^{-Vq/kT} \tag{4.3}$$

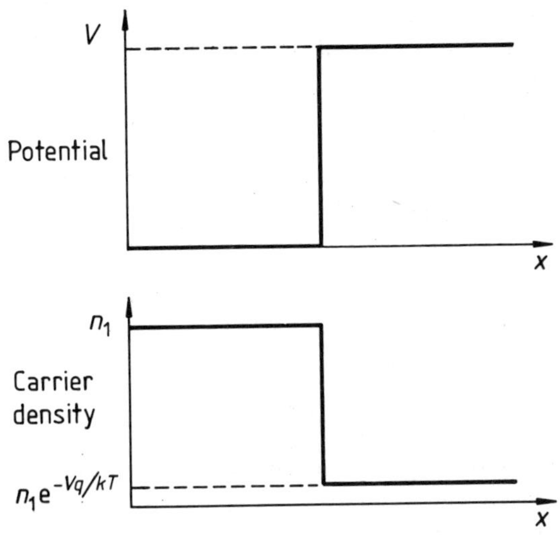

Fig. 4.4 Carrier population on either side of a potential step.

Note that Equation 4.3 tells us nothing about the absolute values of the densities n_1 and n_2 on either side of the potential step, only what their ratio must be at a temperature T (Fig. 4.4).

Accepting these three relationships will allow a quantitative assessment of the junction characteristics.

4.1.2 Current–Voltage Relationship

We have seen that with no applied voltage, or bias, two equal and opposite currents flow in the diode. A diffusion current J_D carries electrons from the n-type material across the junction into the p-type region and a drift current J_S flows in the opposite direction. We can write the total current J_T as

$$J_T = J_D - J_S$$

If no bias is applied to the junction the total current is zero and so

$$J_T = J_D - J_S = 0 \qquad (4.4)$$

If now the carrier densities on either side of the junction are disturbed by applying a bias voltage V as suggested by Equation 4.3 the diffusion current J_D will be altered. Suppose J_D increases to $J_D + \Delta J_D$ then, from Equation 4.4, the net current through the diode is given by

$$J = J_D + \Delta J_D - J_S = \Delta J_D \qquad (4.5)$$

It is seen from Equation 4.5 that to express the diode current as a function of applied voltage it is necessary to compute only the *change* in diffusion current that this voltage produces. This will be done by considering the extra diffusive flow of electrons into some arbitrary volume of the junction and equating it to the extra recombination rate there, since nowhere in the junction may carrier densities increase or decrease once the dynamic equilibrium is established.

In that part of the depletion region lying between the planes x and $x + dx$ we have (Fig. 4.5):

$$\text{Extra diffusion into region at } x = -D\frac{d(\Delta n)}{dx}$$

where Δn is the increase in electron density due to the applied bias voltage. The extra diffusion out of the region at $x + dx$ is

$$-\left(D\frac{d(\Delta n)}{dx} + \frac{d}{dx}\left(D\frac{d(\Delta n)}{dx}\right)dx\right)$$

$$= -\left(D\frac{d(\Delta n)}{dx} + D\frac{d^2(\Delta n)}{dx^2}\right)dx$$

$$\text{Extra recombination rate} = \frac{\Delta n}{\tau}dx$$

The equilibrium condition requires:

$$\text{Net diffusion in} = \text{Recombination}$$

$$D\frac{d^2(\Delta n)}{dx^2} = \frac{\Delta n}{\tau}$$

The solution to this equation is

$$\Delta n = K\,e^{-x/\sqrt{(\tau D)}} \qquad (4.6)$$

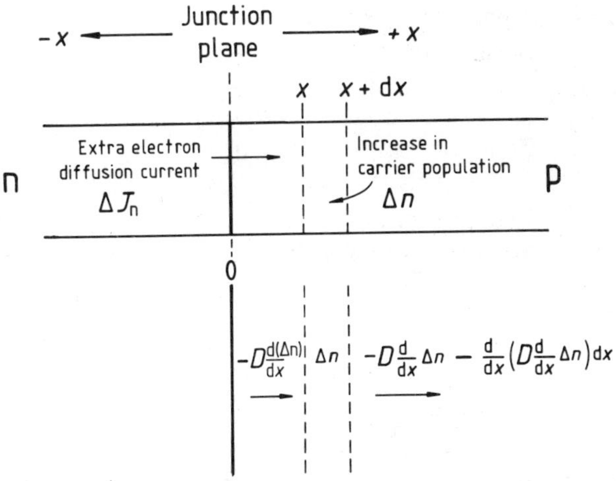

Fig. 4.5 Diffusion current and carrier population changes in biased junction.

The constant of integration K can be evaluated by considering the carrier density at $x = 0$. Before biasing this was n_p. After biasing it must, by Equation 4.3 increase to $n_p e^{qV/kT}$.

Hence at $x = 0$

$$\Delta n = n_p e^{qV/kT} - n_p$$

Putting this condition into Equation 4.6 gives

$$K = n_p(e^{qV/kT} - 1)$$

The explicit value for Δn as a function of x thus becomes

$$\Delta n = n_p(e^{qV/kT} - 1)\, e^{-x/\surd(\tau D)} \tag{4.7}$$

Knowing this, the extra electron current diffusing across a plane at x can be determined using Equation 4.1.

$$\text{Electron current} = \frac{d\,\Delta n}{dt} = -D\frac{d(\Delta n)}{dx}$$

$$= n_p\sqrt{\frac{D}{\tau}}\,(e^{qV/kT} - 1)\, e^{-x/\surd(\tau D)}$$

Since each electron carries a charge q the electrical current ΔJ_n is:

$$\Delta J_n = qn_p\sqrt{\frac{D_n}{\tau_n}}\,(e^{qV/kT} - 1)\, e^{-x/\surd(\tau_n D_n)} \tag{4.8}$$

A similar analysis gives for ΔJ_p the hole current:

$$\Delta J_p = qp_n\sqrt{\frac{D_p}{\tau_p}}\,(e^{qV/kT} - 1)\, e^{x/\surd(\tau_p D_p)} \tag{4.9}$$

Suffixes n and p are introduced to differentiate the electron and hole diffusion coefficients D_n and D_p and recombination times τ_n and τ_p.

The total diode current $\Delta J_n + \Delta J_p$ is obtained by adding Equations 4.8 and 4.9

$$J = \Delta J_n + \Delta J_p = \left(\frac{q}{\sqrt{\tau_n}} n_p\sqrt{D_n}\, e^{-x/\surd(\tau_n D_n)}\right.$$

$$\left. + \frac{q}{\sqrt{\tau_p}} p_n\sqrt{D_p}\, e^{x/\surd\tau_p D_p}\right)(e^{qV/kT} - 1)$$

Since the total current cannot depend on x we may give x any convenient value less than the junction width and choose to take $x = 0$, whence

$$J = q\left(n_p\sqrt{\frac{D_n}{\tau_n}} + p_n\sqrt{\frac{D_p}{\tau_p}}\right)(e^{qV/kT} - 1) \tag{4.10}$$

If a very large negative bias is applied to a diode the diffusion current will be stopped almost completely leaving only the saturation or drift current J_S flowing. The value of J_S is obtained from Equation 4.9 by making V large and negative when

$$J \to -q\left(n_p\sqrt{\frac{D_n}{\tau_n}} + p_n\sqrt{\frac{D_p}{\tau_p}}\right)$$

hence the magnitude of J_S can be written

$$J_S = q\left(n_p\sqrt{\frac{D_n}{\tau_n}} + p_n\sqrt{\frac{D_p}{\tau_p}}\right)$$

making Equation 4.9

$$J = J_S(e^{qV/kT} - 1) \tag{4.11}$$

The shape of this J/V relationship is shown in Fig. 4.2.

The effective resistance of such a circuit element is of interest. Clearly the value of V/J varies with V but the simple ratio is usually of less concern to the circuit designer than is dV/dJ, the slope of the V/J curve. Knowledge of dV/dJ allows the calculation of the small change in V which a small change in J would bring about, or vice versa. If both sides of Equation 4.11 are differentiated with respect to V then it is seen that

$$\frac{dV}{dJ} = \frac{kT}{J_S q}\, e^{-qV/kT} \tag{4.12}$$

This is known as the small-signal or incremental resistance and it is seen that with a large negative bias it becomes very large, at zero bias it has the value $kT/J_S q$ and with large forward bias it becomes so small that it is usually obscured by contact imperfections in the device.

4.1.3 Junction Width and Capacity

There is a narrow region on the n side of the junction whence all the electrons from the donor centres have diffused into the p-type material on the other side. If there were originally N_d donor centres per m^3 then the electrical charge density will be approximately qN_d coulomb/m^3 in this depleted region. A similar region on the p side of the junction will have an approximate charge density $-qN_a$ coulomb/m^3 if the p-type material originally had N_a acceptor centres per m^3. We require to find how far back into the bulk semiconductor these DEPLETION REGIONS extend. Using Equation 4.2 and referring to Fig. 4.6 we write

$$\frac{d^2V}{dx^2} = -\frac{qN_d}{\varepsilon}$$

for the n-type depletion region.

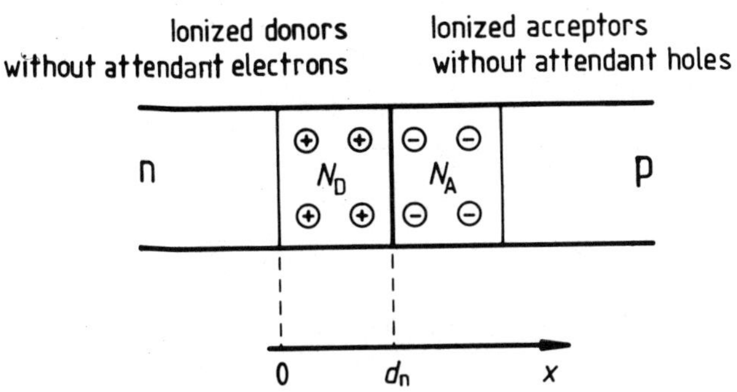

Fig. 4.6 Depletion regions in junction.

Integrating this equation gives

$$\frac{dV}{dx} = -\frac{qN_d}{\varepsilon}x + K$$

At the edge of the region $dV/dx = 0$ and $x = 0$. Therefore $K = 0$

$$\frac{dV}{dx} = -\frac{qN_d x}{\varepsilon}$$

$$V = -\frac{qN_d x^2}{2\varepsilon} + L$$

When $x = 0$, $V = 0$ and therefore $L = 0$

$$V = -\frac{qN_d x^2}{2\varepsilon}$$

Let the potential at the centre of the junction, i.e. when $x = d_n$, be V^1.

$$V^1 = -\frac{qN_d d_n^{\,2}}{2\varepsilon} \qquad (4.13)$$

A similar exercise for the depletion region in the p side gives

$$V_d - V^1 = +\frac{qN_a d_p^{\,2}}{2\varepsilon} \qquad (4.14)$$

Since the total charge in the junction must be zero

$$d_n N_d = d_p N_a \qquad (4.15)$$

From Equations 4.12, 4.13 and 4.14 we obtain:

$$d_n + d_p = \left[\frac{N_d + N_a}{N_d N_a} 2 \frac{\varepsilon V_d}{q}\right]^{1/2}$$

The junction may be regarded as a capacitor having a dielectric slab of thickness $d_n + d_p$ with a permittivity of ε. The capacity C' will be given by

$$C' = \frac{\varepsilon A}{(d_n + d_p)}$$

Where A is the junction area. For unit area

$$C = \sqrt{\frac{N_d N_a q\varepsilon}{2(N_d + N_a)V_d}} \qquad (4.16)$$

It should be noted that V_d is the total voltage difference across the junction. This will be increased by a reverse bias causing $(d_n + d_p)$ to increase, and hence decrease the value of C. A forward bias will have the opposite effect.

This analysis applies to an ABRUPT JUNCTION wherein a constant value of N_a changes abruptly to a constant value of N_d. It is possible for the junction to be manufactured in such a way that the impurity density changes smoothly from zero at the junction centre to N_d in the n-type and to N_a in the p-type side (Fig. 4.7a). The distribution can be simply written as

$$N_{\text{IMPURITY}} = -ax$$

where a is the gradient of the impurity distribution. The negative sign makes the impurity density positive

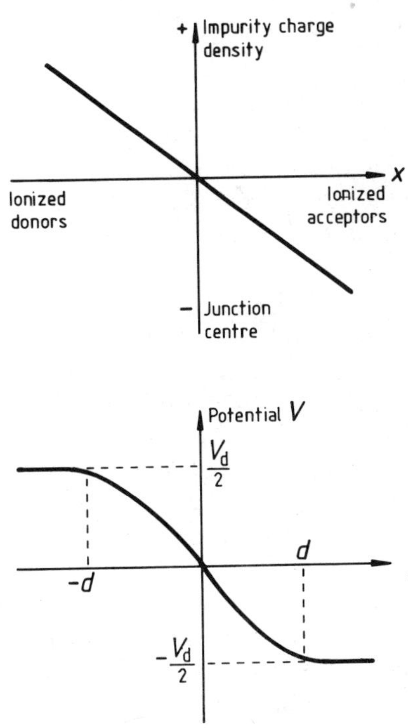

Fig. 4.7 Potential and charge density in graded junction.

to represent ionized donor centres when x is positive. Following an analysis similar to that used for the abrupt junction we can write

$$\frac{d^2V}{dx^2} = +\frac{qax}{\varepsilon}$$

$$\frac{dV}{dx} = \frac{qax^2}{2\varepsilon} + K$$

At the junction boundaries $dV/dx = 0$ and $x = \pm d$ (Fig. 4.7b).

$$K = -\frac{qad^2}{2\varepsilon}$$

$$\frac{dV}{dx} = \frac{qa}{2\varepsilon}(x^2 - d^2)$$

$$V = \frac{qa}{2\varepsilon}\left(\frac{x^3}{3} - d^2x\right) + L \qquad (4.17)$$

At the junction boundaries

$$x = +d \text{ with } V = -V_d/2$$

and

$$x = -d \text{ with } V = +V_d/2.$$

$$-\frac{V_d}{2} = -\frac{qad^3}{3\varepsilon} + L$$

$$\frac{V_d}{2} = \frac{qad^3}{3\varepsilon} + L$$

Subtracting gives

$$V_d = \frac{2qad^3}{3\varepsilon}$$

or

$$2d = \left(\frac{12\varepsilon V_d}{qa}\right)^{1/3}$$

The capacity of the GRADED JUNCTION follows as

$$C = \frac{\varepsilon}{\left(\frac{12\varepsilon V_d}{qa}\right)^{1/3}} = \left(\frac{qa\varepsilon^2}{12V_d}\right)^{1/3} \qquad (4.18)$$

The variation of capacity with total diode voltage is shown in Fig. 4.8 for the two types of junction. Note that the total diode voltage is equal to the zero bias voltage minus the applied voltage.

$$V_c = V_0 - V_A$$

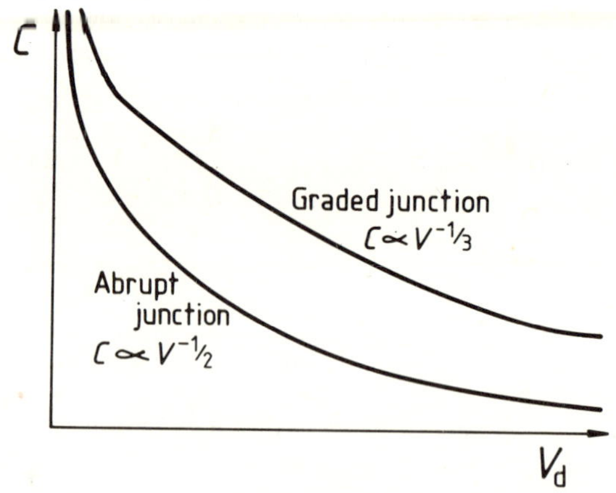

Fig. 4.8 Junction capacity as a function of total voltage across the junction.

4.2 SCHOTTKY JUNCTIONS

4.2.1 Introductory

If any two conducting materials having the same temperature are in electrical contact, the fermi levels must be continuous across the junction. This follows because the fermi level represents a form of averaged energy for the electrons in the electron gas in each material. If either energy exceeded the other, carriers passing from the more energetic to the less would raise the temperature of the electron gas in the latter. This cannot, by definition, occur in equilibrium and so the mean energy and fermi level of each material must have the same absolute value. This is the only occasion upon which it is necessary to consider, even superficially, the absolute value of the fermi level. Generally we are concerned with its relationship to the conduction band, valence band or some impurity centre energy.

To be specific we will now examine a junction found between a metal and an n-type semiconductor. The interface between the materials will normally contain many imperfections which act as trapping centres. If they trap holes then the region of semiconductor near the junction will be denuded of holes and, since the fermi level must be flat, Equation 2.18a shows that the bands must bend downwards at the junction (Fig. 4.9a). Electrons can now flow easily between the metal and the conductor and the junction is hence described as an OHMIC CONTACT for electrons. The analogous ohmic contact for holes follows similarly with a p-type semiconductor and electron traps.

Had the first junction envisaged given rise to electron traps at the interface, the bands would have bent

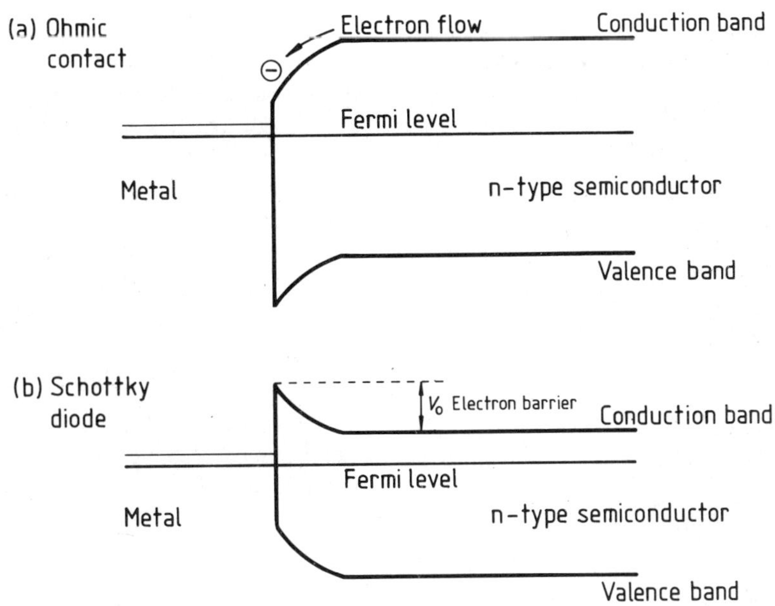

Fig. 4.9 Metal-semiconductor junctions.

upwards at the junction and formed a barrier to the passage of electrons (Fig. 4.9b). Such a barrier is known as a SCHOTTKY JUNCTION and has characteristics similar to the p–n junctions discussed in Section 4.1.

4.2.2 Current-Voltage Relationship

The current-voltage relationship in the Schottky junction can be obtained by an analysis corresponding to that used for the p–n junction in Section 4.1.2. It is necessary to replace the diffusion current J_D by an emission current J_E carried by electrons emitted over the potential barrier V_0 formed by the bent bands. It is in fact a solid-state analogy to the current emitted by a hot cathode in a thermionic valve. If there is an electron density n in the body of the semiconductor, then Equation 4.3 shows that a density of electrons $n\,e^{-V_0/kT}$ will have sufficient energy to surmount the potential barrier. They will have a thermal velocity of $\sqrt{(2kT/m_n)}$ but, in their random motion, roughly half of them will be travelling towards the metal and so the emission current density will be

$$J_E = \tfrac{1}{2}\,(\text{electron charge} \times \text{density} \times \text{velocity})$$

$$= \frac{1}{2} q \times n\,e^{-V_0/kT} \times \sqrt{\left(\frac{2kT}{m_n}\right)}$$

$$= q\sqrt{\left(\frac{kT}{2m_n}\right)}\,n\,e^{-V_0/kT}$$

If a potential V_A is applied in such a manner as to reduce the potential barrier to $(V_0 - V_A)$ then $J_E + \Delta J_E$ where

$$J_E + \Delta J_E = q\sqrt{\left(\frac{kT}{2m_n}\right)}\,n\,e^{-(V_0 - V_A)/kT}$$

whence

$$\Delta J_E = q\sqrt{\left(\frac{kT}{2m_n}\right)}\,n\,e^{-V_0/kT}(1 - e^{V_A/kT})$$

As with the p–n junction this is the external current flowing when a bias voltage V_A is applied

$$J = q\sqrt{\left(\frac{kT}{2m_n}\right)}\,n\,e^{-V_0/kT}(e^{V_A/kT} - 1) \qquad (4.19)$$

or

$$J = |J_s|(e^{V_A/kT} - 1)$$

where J_s is the saturation current obtained by making V_A large and negative.

4.2.3 Junction Width and Capacity

If the semiconductor has been made n-type by the addition of a density N_D of donor impurities then the charge density in the electron denuded region will be $+qN_D$. Again using Equation 4.2 we have

$$\frac{d^2V}{dx^2} = -\frac{qN_D}{\varepsilon}$$

This equation has previously been solved to yield Equation 4.13 which can be rewritten here as

$$V_0 - V_A = \frac{qN_D d^2}{2\varepsilon}$$

The depletion layer thickness d can now be used to obtain C' the junction capacity:

$$C' = \frac{\varepsilon A}{d} = \sqrt{\left(\frac{\varepsilon q N_D}{2(V_0 - V_A)}\right)} A$$

for a junction of area A. The capacity per unit area C is thus

$$C = \sqrt{\left(\frac{\varepsilon q N_D}{2(V_0 - V_A)}\right)} \qquad (4.20)$$

This should be compared with Equation 4.16.

4.3 THE BIPOLAR TRANSISTOR

4.3.1 Qualitative Description

The original transistors used the charge carriers in both p- and n-type material and are described as 'bipolar' to differentiate them from the unipolar, or field effect, devices that are discussed in Chapter 6. In essence the bipolar transistor consists of two p–n junctions set close together in a single block of material (Fig. 4.10). One of the diodes (the emitter) is forward biased and the current flowing across it into the base material diffuses across to the second p–n junction (the collector) which is reverse biased. When the carriers forming this current reach the high-field region of the collector they are driven across it and through an external circuit. Now, in a well designed, well constructed transistor practically the whole of the emitter current reaches the collector.

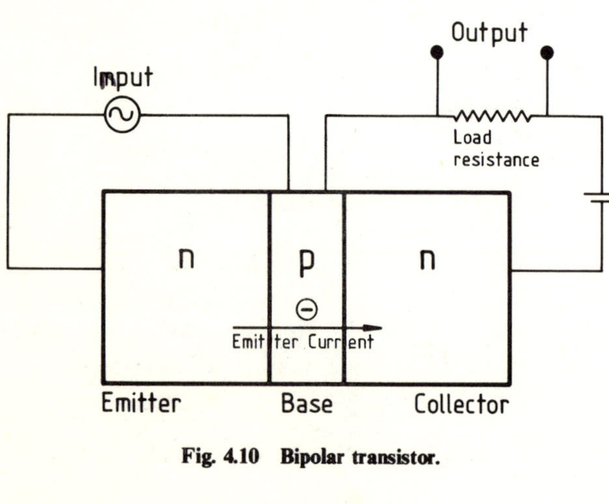

Fig. 4.10 Bipolar transistor.

Since the currents flowing in the emitter and collector circuits are almost equal, the power in each circuit, I^2R, is proportional to the circuit resistance. The forward-biased emitter has a low resistance while the reverse-biased collector resistance is high and so allows a large load resistance R_L in the circuit. Thus a small power variation in the collector circuit yields a large power variation in the emitter circuit. Amplification has taken place.

4.3.2 Amplification Magnitude

The power in the input circuit of the transistor, P_{IN}, can be expressed in terms of the emitter current and voltage I_E and V_E:

$$P_{IN} = I_E V_E$$

Variations in P_{IN} are obtained by differentiation:

$$dP_{IN} = dI_E V_E + dV_E I_E$$

which becomes

$$dP_{IN} = \frac{kT}{q} dI_E + \frac{kT}{q} \ln\left(\frac{I_E}{I_{EO}}\right) dI_E \qquad (4.21)$$

by using the emitter diode equation

$$I_E \doteq I_{EO}(e^{qV_E/kT})$$

The power in the load resistance R_L, written as P_{OUT}, is

$$P_{OUT} = R_L I_C^2$$

where I_C is the collector diode current. From this

$$dP_{OUT} = 2I_C R_L dI_C \qquad (4.22)$$

The amplification A is defined as

$$\frac{dP_{OUT}}{dP_{IN}}$$

$$A = \frac{dP_{OUT}}{dP_{IN}} = \frac{2I_C R_L dI_C}{\frac{kT}{q}\left[1 + \ln\left(\frac{I_E}{I_{EO}}\right)\right] dI_E}$$

As

$$I_C \approx I_E = I$$

$$A = \frac{2IR_L}{\frac{kT}{q}\left(1 + \ln\left(\frac{I}{I_{EO}}\right)\right)} \qquad (4.23)$$

The value of the load resistor R_L must not greatly exceed the reverse-biased resistance of the collector junction or the output current will flow internally across the junction rather than through the external circuit.

4.4 TUNNEL DIODES

4.4.1 Tunnelling

If a 'classical' particle, that is to say one obeying Newtonian mechanics, is projected at an obstacle, it will surmount it providing it has sufficient energy, or it will bounce back if it has not. Consider a marble with a kinetic energy K rolling towards a mound of height H and thickness t (Fig. 4.11). The mound presents an energetic barrier of magnitude mgH where m is the mass of the marble and g the acceleration due to gravity. Writing the energy mgH as E, then the marble will surmount the obstacle and continue on its way if K is greater than E. If K is less than E the marble will rise part way up the mound and then roll back with its kinetic energy reversed in sign. To sum the situation:

If $K > E$ the marble is transmitted by the obstacle
If $K < E$ the marble is reflected by the obstacle

Now for a very small particle such as an electron the situation is much less clear cut. Whatever the relative values of E and K there is a finite chance that the electron will penetrate the barrier. The probability that it will do so will become less as E increases with respect to K, and for any given values of E and K the probability decreases as t increases.

The general trend of the behaviour can be anticipated by considering the Uncertainty Principle discussed in Chapter 2. If the position of the electron is known precisely by making t very small then its energy is very uncertain and may be big enough to overcome the obstacle. If t is large then the position in the barrier is uncertain and the electron energy more precisely known, giving less chance of surmounting the barrier.

The uncertainty in the momentum p of the electron is given by the Uncertainty Principle as

$$\Delta p \times t \geqslant h$$

Now $p = \sqrt{(2mK)}$ hence $\Delta p = \sqrt{(m/2K)}\,\Delta K$. Putting this value in the uncertainty unequality gives a minimum

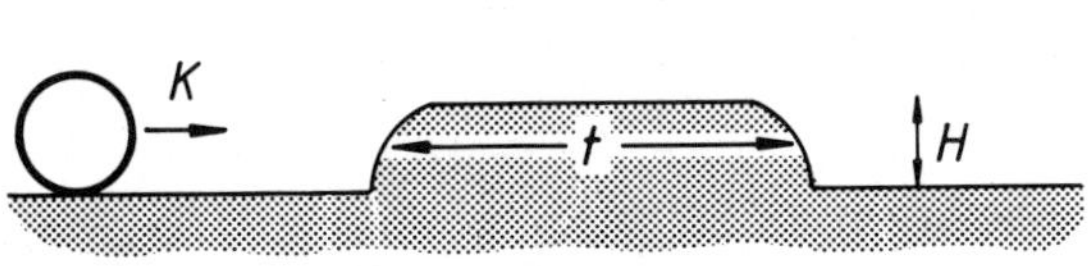

Fig. 4.11 Marble and obstacle.

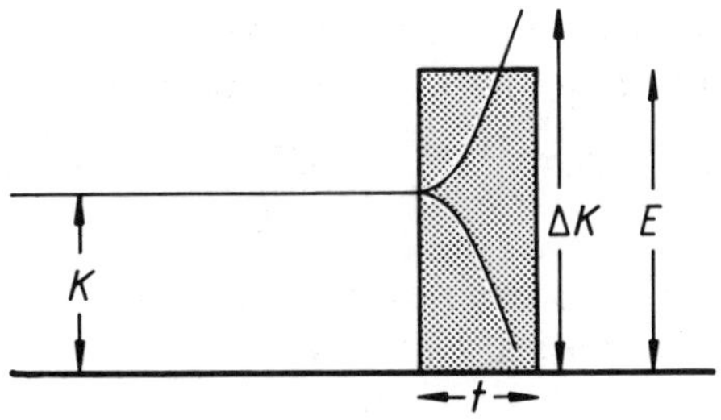

Fig. 4.12 Energy uncertainty for electron at barrier.

value for ΔK, the uncertainty in the kinetic energy, as

$$\Delta K \geqslant \frac{h}{t}\sqrt{\left(\frac{K}{2m}\right)} \tag{4.24}$$

From the diagram (Fig. 4.12) we might expect P, the probability of transmission through the barrier, to depend on the amount by which the value of $K + (\Delta K/2)$ exceeds the barrier height E.

To a good approximation

$$P = e^{-(2t/h)\sqrt{[m(E-K)]}} \tag{4.25}$$

4.4.2 The Tunnel Diode

If the impurity densities in a p–n junction are very large then, as Section 4.1.3 shows, the depletion region will become very small. In the limit it may become so thin that carriers can pass right through it by the tunneling process just considered. A junction operating on this principle is known as an ESAKI or TUNNEL DIODE. Fig. 4.13 shows the appearance of the band structure.

The impurity densities on both sides of the junction are assumed to be so great that the fermi level has

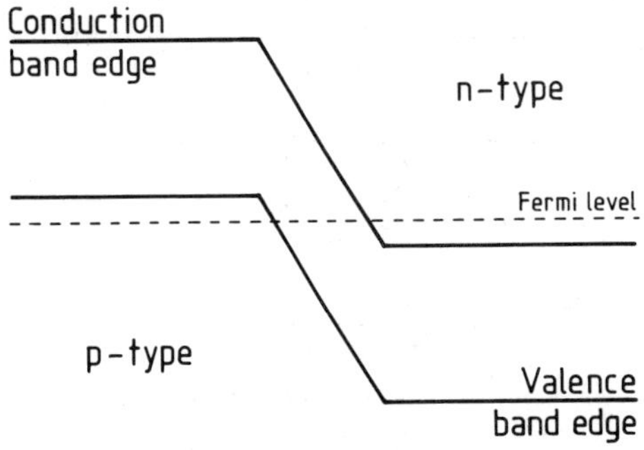

Fig. 4.13 Energy bonds in tunnel diode.

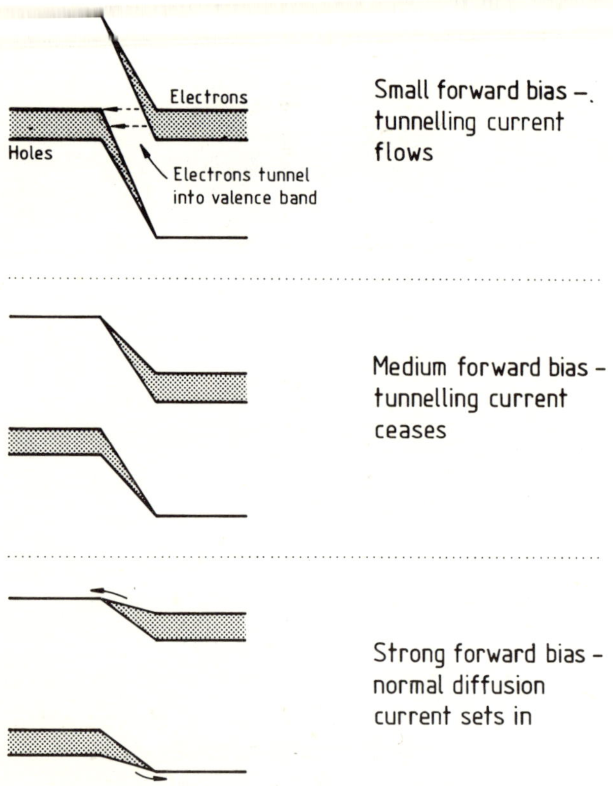

Fig. 4.14 Tunnel diode under various bias conditions.

moved right into the valence band in the p-type material and into the conduction band in the n-type. Both are thus degenerate or near metallic. If now a small forward bias is applied, electrons in the bottom of the conduction band will tunnel through the depletion region into the empty spaces, or holes, in the bottom of the valence band (Fig. 4.14).

With a medium forward bias the bottom of the conduction band rises above the top of the valence band and the tunneling current stops as there are no slots, or holes, into which the electrons can tunnel. As the bias is increased still further the conventional diffusion current becomes significant and the sum of these two currents is shown as a function of voltage in Fig. 4.15.

There are two particular characteristics of this device which make it of interest to the circuit engineer. Since tunneling takes place at the speed of light rather than at a velocity equal to the mobility multiplied by the electric field as in normal conduction, the diode responds very quickly to signals applied to it. A response time of 5×10^{-11} s has been reported. A second useful effect arises from the fact that there is a part of the current-voltage curve where an increase in voltage actually decreases the current; that is, the resistance of the device shows a negative value. Now any resonant circuit will oscillate if a pulse of current is applied to it, but because of the normally positive value of the circuit resistance the train of oscillations will decrease and die out (Fig. 4.16a).

If the resistance is negative the oscillators will not decrease but will be maintained at a steady amplitude (Fig. 4.16b). They will not, as one might suppose, increase without limit as the negative resistance, in this case a tunnel diode, is itself limited to currents lying between the peak and trough of the diode characteristic (Fig. 4.15).

The tunnel diode can therefore be used in fast switching circuits and in oscillators, these usually working at about 10^{10} Hz and at low power.

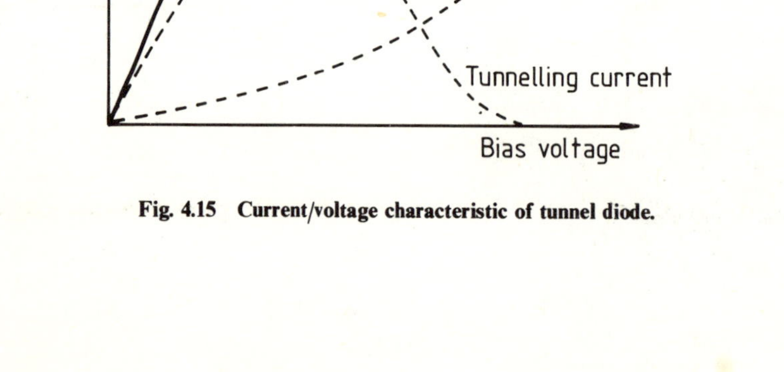

Fig. 4.15 Current/voltage characteristic of tunnel diode.

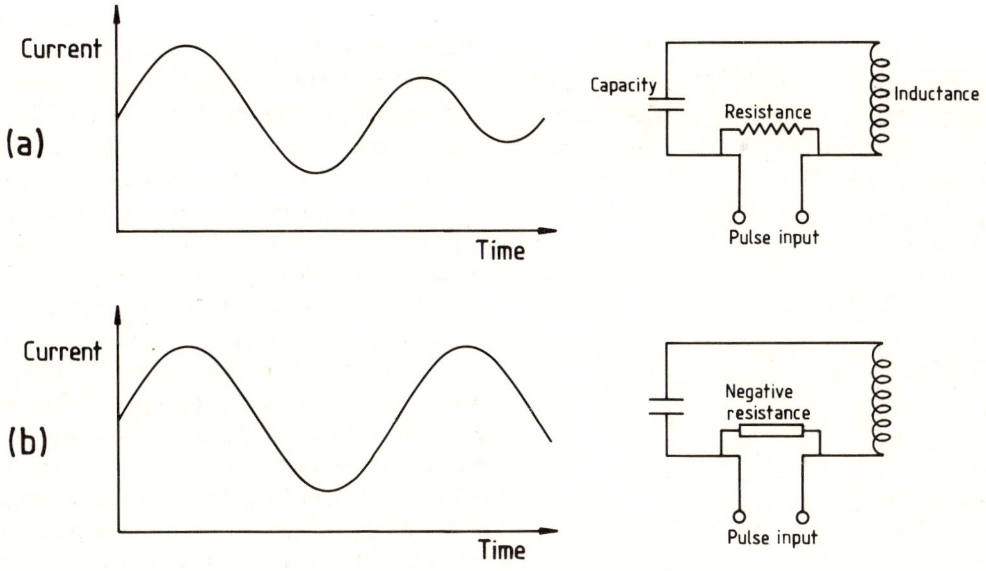

Fig. 4.16 Oscillatory circuit behaviour.

4.5 DIODE BREAKDOWN

4.5.1 Avalanche Breakdown

If a strong reverse bias is applied to a diode, very high fields will exist in the depletion layer. These can be so great that the free electrons are accelerated to velocities large enough to knock fixed electrons from their parent atoms, thus increasing the number of free carriers. These carriers in their turn will release further fixed electrons and a multiplication, or avalanching effect, will take place. This increase in the number of free carriers, and hence the conductivity, takes place abruptly at a given bias value called the BREAKDOWN voltage. The current-voltage curve shown in Fig. 4.2 can now be extended to include this region as shown in Fig. 4.17.

Since this effect relies on the number of free carriers being substantially increased by collisions, it will be more difficult to produce at higher temperatures, when there are more naturally occurring carriers available. The bias voltage needed for breakdown increases by about 0.5 mV for every °C increase in temperature.

4.5.2 Zener Breakdown

The avalanche mechanism may not set in until as much as 100 V reverse bias is applied. At reverse biases below about 6 V a different effect known as Zener breakdown can set in. In this, carriers are freed from their parent atoms by the direct force of the electric field which will amount to millions of volts per metre. The very precise voltage at which the Zener breakdown takes place makes these junctions useful for voltage limiting circuits. Connected across the output of a voltage source they will form an effective short circuit should the voltage exceed the breakdown value.

Unlike the avalanche effect, Zener breakdown occurs at lower voltages as the temperature increases. This is because the band gap itself, across which the carriers must be forced, decreases with increasing temperature. The breakdown bias voltage increases by a few tenths of a millivolt for every °C increase in temperature.

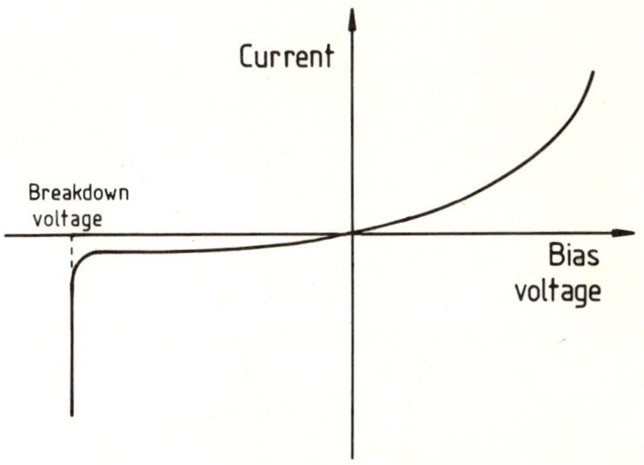

Fig. 4.17 Diode characteristic showing breakdown region.

APOLOGIA

This chapter has taken an orthodox, if simplified, attitude to junction analysis, since such a procedure will be necessary for readers liable to formal examination and should at least be considered by others. It must however be remarked that observed junction behaviour is grossly at variance with established theory save in some junctions made from simple elemental semiconductors. Various piecemeal modifications are available in the literature but the subject at present awaits an original and coherent mind.

PROBLEMS

For values of required constants, see Problems to Chapter 1.

4.1 Light falling one one end of a bar of semiconducting material generates free electrons at a rate of 10^{11} s^{-1}. If all the electrons diffuse to the far end of the bar, where they recombine instantaneously, what is their density at the centre of the bar? The bar has a length of 0.1 m and a cross-section of 10^{-6} m^2 with a diffusion coefficient for free electrons of 5×10^{-3} m^2 s^{-1}.

4.2 When a bias of + 0.5 V is applied to an abrupt junction diode at a temperature of 300 °K, a current of 10^{-4} A flows. If the bias is reduced to 0.3 V, what current will then flow?

4.3 What is the incremental resistance of the diode in Question 4.2 at the lower bias voltage?

4.4 The capacity of an abrupt junction diode at a bias of + 1 V is 5×10^{-12} F and at a bias of − 1 V is 1.8×10^{-12} F. What is the capacity at zero bias?

4.5 What is the voltage step across the unbiased junction in Question 4.4?

4.6 What values would the zero bias capacity and the voltage step across the unbiased junction have had if the diode in Question 4.5 had been graded rather than abrupt?

4.7 Show that in a 'well designed, well constructed' bipolar transistor the amplification increases as the emitter current increases beyond the value I_{EO}.

4.8 A free electron approaches a potential barrier of 5 eV after being accelerated through a potential rise of 1 V. If the barrier thickness is 9.3×10^{-10} m, what is the probability that the electron will be reflected from the barrier?

4.9 An abrupt p–n junction made from material with an energy gap of 1 eV has a depletion layer of 10^{-6} m when unbiased. If avalanche breakdown sets in when a reverse bias voltage of 8 V is applied, what is the approximate mean free path of the electrons in the junction?

CHAPTER 5

Optical, Thermal and Magnetic Effects

The number of free electrons in a semiconductor can be greatly increased by heating or shining light upon it. These effects not only allow its use as a thermometer or light detector but enable fundamental measurements of the material properties to be made.

If the free electrons are drifting under the influence of an electric field and a magnetic field is then applied also, the direction of drift will be altered. Precise measurements of this change of direction will allow the actual numbers and velocities of the free electrons to be calculated.

These thermal, optical and magnetic effects, and their use in devices, are dealt with in this chapter.

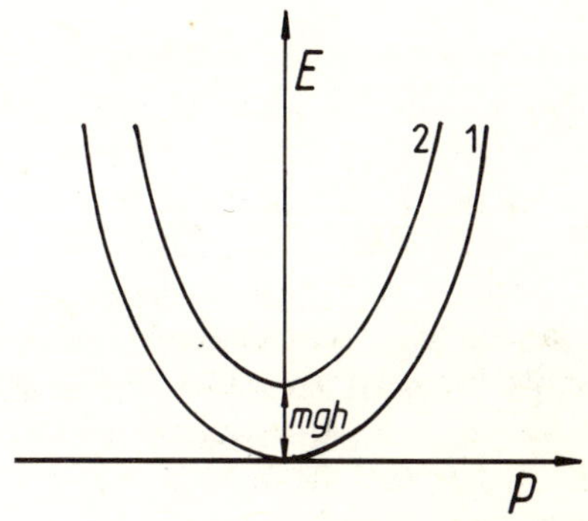

Fig. 5.1 Energy-momentum relationship for Newtonian body.

5.1 OPTICAL EFFECTS

5.1.1 Introductory

To understand the manner in which photons, the packets of light energy, interact with electrons in a semiconductor, first consider the energy-momentum relationship of a classic Newtonian particle. The marble used in tunnelin considerations in Chapter 4 will serve. The kinetic energy E of the marble when moving with a velocity V is

$$E = \tfrac{1}{2}mV^2 \qquad (5.1)$$

and the momentum p is

$$p = mV \qquad (5.2)$$

From Equations 5.1 and 5.2

$$E = \frac{1}{2}\frac{p^2}{m} \qquad (5.3)$$

A graph of this relationship is shown in curve 1 of Fig. 5.1.

The negative values of p indicate a velocity in the opposite direction to that implied by the positive p quadrant. If the marble is now raised through a height h then a potential energy mgh is added to all parts of the curve, g being the acceleration due to gravity. The curve now shifts to position 2.

The energy-momentum curves of an electron within a semiconductor are much more complicated because the effective mass is not only different from that of a free electron, as we have seen, but actually varies with the energy also. This effect arises from the interaction of the electron with the varying potential inside the semiconductor. A typical E/p curve for an electron in the conduction band could look like curve C in Fig. 5.2. A corresponding curve for an electron in the valence band is shown at V. The separation between the lowest energy in the conduction band, shown at a, and the highest energy in the valence band, at b, is the energy gap of the material. This diagram should be compared with Fig. 2.1, where the lowest conduction band energy and the highest valence band energy are drawn against position in the semiconductor body ignoring momentum.

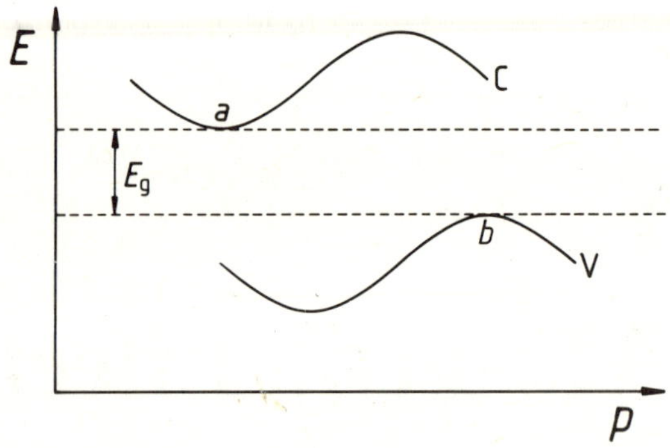

Fig. 5.2 Energy-momentum curves for electron in semiconductor.

We must now see how an energetic particle can be used to excite an electron from the top of the valence band to the bottom of the conduction band to give a free hole and electron. If the particle were our faithful Newtonian marble then a transition on an energy-momentum diagram would take it from the origin to a point such as A in Fig. 5.3, with energy and momentum changes of $\frac{1}{2}mV^2$ and mV respectively. The angle θ that this transition makes on the diagram is given by

$$\tan\theta = \tfrac{1}{2}mV^2/mV = \frac{V}{2}$$

The electrons in a semiconductor interact not with marbles but with photons and phonons. The photon, as we have said, may be regarded as a parcel of light or radiation energy. Its energy E is related to the wavelength of the radiation λ by

$$E = \frac{1.24}{\lambda} \tag{5.2}$$

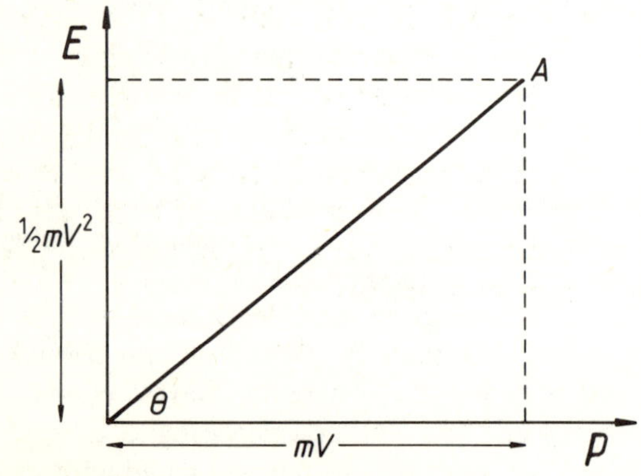

Fig. 5.3 Energy-momentum angle.

if E is measured in eV and λ in microns. Since it travels with the velocity of light its tan θ value is large and θ is almost 90°, giving a practically vertical transition on Fig. 5.3. This model for the large-energy little-momentum behaviour of the photon must be treated with caution as it contains an implicit suggestion that photons have normal, even if very small, masses.

The photon is a packet of heat energy, always available in a body at any finite temperature, which shows itself in a solid as a travelling vibration of the atoms in the lattice. It is in fact like a sound wave of very high frequency. Compared with the velocity of light the velocity of sound in a solid is small and so transitions caused by phonons have much smaller θ values on the energy-momentum diagram.

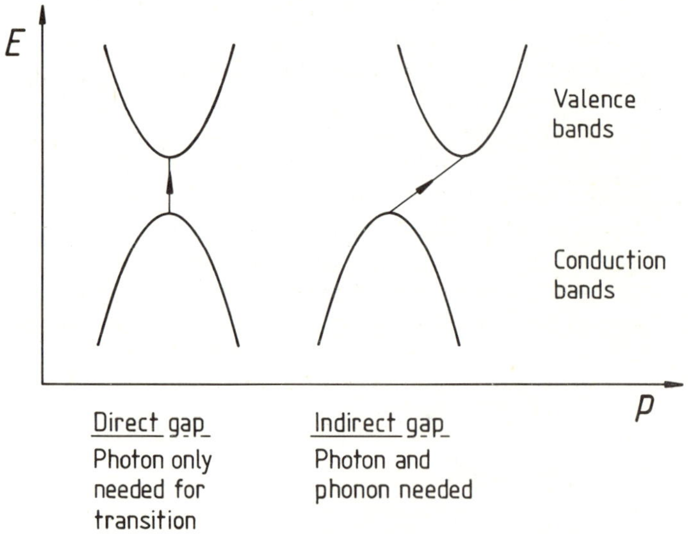

Fig. 5.4 Transitions in direct and indirect gap materials.

Referring back to Fig. 5.2 we now see that for an electron to be excited from a to b will require the energy of a photon together with the momentum of a phonon. The semiconductor suggested in Fig. 5.2, where a and b are offset on the momentum axis, is described as an INDIRECT GAP material. It is evident that transitions are difficult to effect since they require the simultaneous action of a photon and a phonon. If, however, a and b lie vertically above one another, the material is referred to as DIRECT GAP and the transition requires only a photon, clearly a much easier condition to satisfy. These considerations are summarised in Fig. 5.4.

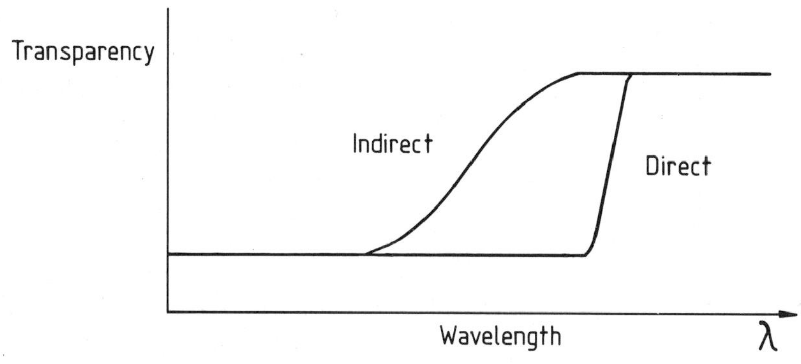

Fig. 5.5 Band edge absorption.

5.1.2 Transparency

If radiation of very long wavelength falls on a semiconductor, the energy in its photons will be small, as Equation 5.2 shows. If this energy is smaller than the band gap, they will not be able to excite electrons from the conduction band to the valence band and will pass straight through the material. It is transparent to this wavelength. If now the wavelength is decreased the photon energy will increase until it is just able to excite electrons across the band gap. The transparency will begin to decrease, rapidly for a direct gap material but less abruptly for indirect transitions, which require the help of any available phonons. Fig. 5.5 shows these effects.

Such curves, measured experimentally, will show both the magnitude of the energy gap and its type; that is, direct or indirect.

If the material contains impurities they will produce a short region of low transparency on the long wavelength side of the transparent-opaque transition. The reason for the limitation on the IMPURITY ABSORPTION region is shown in Fig. 5.6. The impurities exist

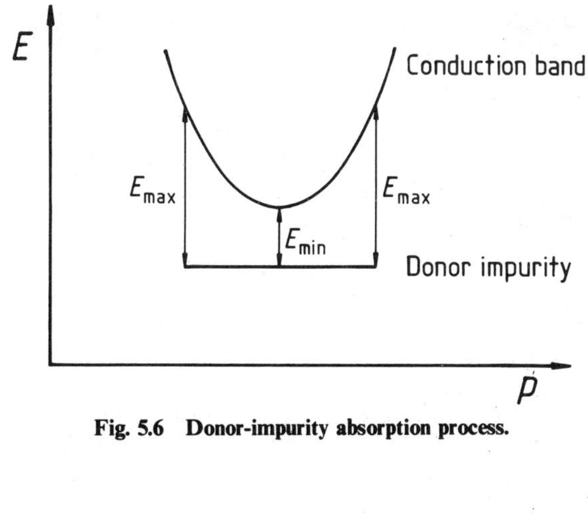

Fig. 5.6 Donor-impurity absorption process.

over only a limited range of momentum values and hence only photons with energies lying between E_{MAX} and E_{MIN} are absorbed.

One more mechanism needs to be considered for a complete picture of semiconductor transparency. This is known as FREE CARRIER ABSORPTION and arises because any free carriers already produced by other processes may be made to oscillate with the frequency of the incident radiation, provided this is low enough, thus absorbing energy. Such low frequencies correspond to very long wavelengths and this gives a slow fall in transparency in the long wavelength region. Fig. 5.7 shows all three absorption regions.

5.1.3 Photoconductive and Photovoltaic Effects

Radiation having the energy (or wavelength) necessary to excite electrons into the conduction band, from either the valence band or impurity centres, will increase the carrier population and hence the conductivity. The magnitude of the effect is quite easily determined. Suppose ϕ photons per second fall on one face of a unit cube of semiconductor containing donor impurities as shown in Fig. 5.8. If each photon is successful in producing a free electron, the rate of production will be ϕ electrons per second. The number of free electrons present will rise by an amount Δn making the rate of recombination, or death rate, increase by $\Delta n/\tau$ where τ is the lifetime of an electron. When the population has reached its new steady value, production and recombination must be equal and so

$$\phi = \frac{\Delta n}{\tau}$$

or

$$\Delta n = \phi\tau \tag{5.3}$$

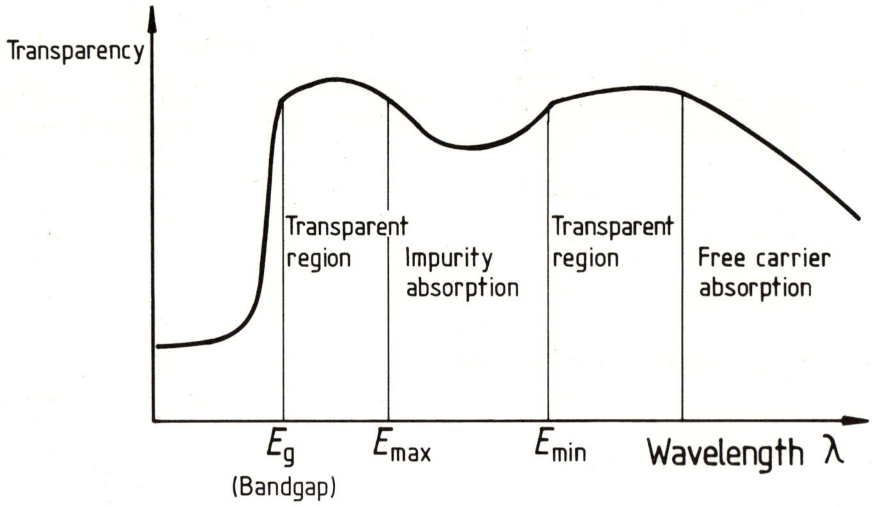

Fig. 5.7 Transparency in semiconductor.

The increase in conductivity $\Delta\sigma$ is then given by

$$\Delta\sigma = \Delta n\mu q$$

$$\Delta\sigma = \phi\tau\mu q \tag{5.4}$$

In general not every photon succeeds in producing an electron. The fraction of effective photons is known as the QUANTUM EFFICIENCY and is generally written as η. The photon flux ϕ thus produces $\phi\eta$ electrons per second making the general form of Equation 5.4:

$$\Delta\sigma = \phi\eta\tau\mu q \tag{5.5}$$

The absolute change in the conductivity is usually less important than the fractional increase. If the undisturbed electron population is n then the conductivity σ will be $n\mu q$. Using this and Equation 5.5 gives

$$\frac{\Delta\sigma}{\sigma} = \frac{\eta\phi\tau}{n} \tag{5.6}$$

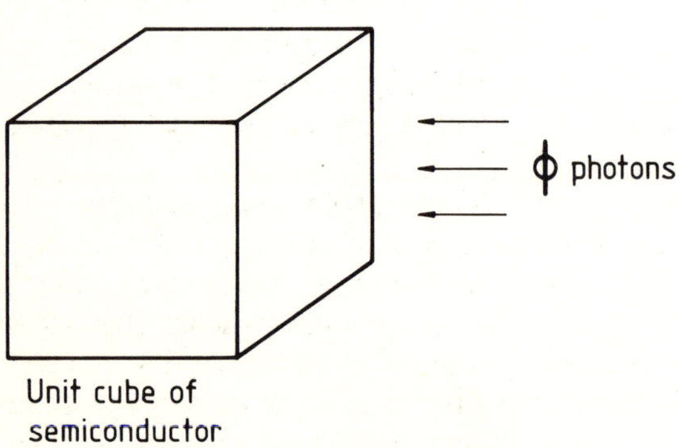

Fig. 5.8 Photoconductive effect.

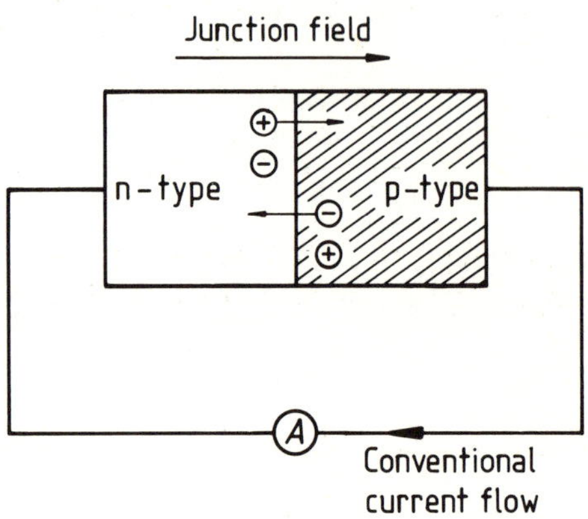

Fig. 5.9 Origin of photovoltaic current.

To observe this change of conductance it would be necessary to have a measuring circuit including a source of electrical energy, say a battery. The photovoltaic effect allows us to dispense with the battery, for it produces its own current.

If electron-hole pairs are liberated by photons near the junction region of a diode, the field will sweep the electrons in the p-type material across the junction and the holes in the n-type region will be swept across in the opposite direction, both contributions being in the reverse, and not the forward, current direction for the junction. An external circuit connected to the diode will experience a current flow due to these charge movements in the junction (Fig. 5.9).

If the circuit is of low resistance, this current will be

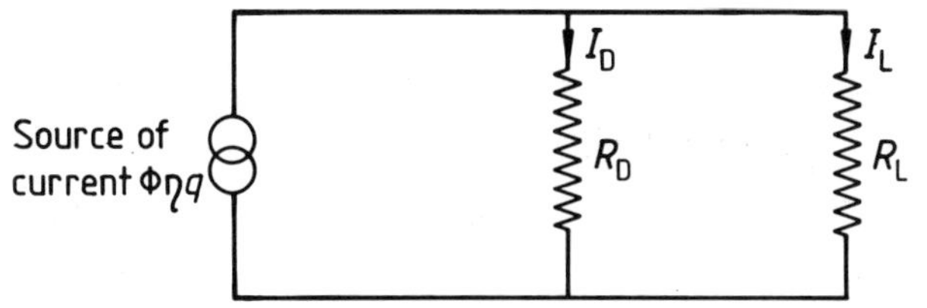

Fig. 5.10 Equivalent circuit of photovoltaic diode.

numerically equal to the rate of excitation of free carriers:

$$\text{Current} = \eta\phi \text{ electron charges per second}$$
$$= \eta\phi q \text{ amps}$$

If a finite resistance R_L is connected across the diode the situation is more complicated. The total current produced by the diode remains as $\phi\eta q$ but it is now divided between R_L and the diode resistance R_D as suggested in the equivalent circuit diagram of Fig. 5.10. The complexity arises because R_D, the diode resistance, is dependent on the current I_D which flows through it.

If R_L is made infinite, that is open-circuited, then the voltage appearing across the diode is obtained by substituting $\phi\eta q$ for the diode current in Equation 4.11 giving

$$\text{Open circuit photovoltage} = \frac{kT}{q} \log_e \left(\frac{\phi\eta q}{I_0} + 1 \right)$$

A typical silicon photocell shows 0.3 V in normal daylight.

5.1.4 Injection Luminescence

In the photovoltaic effect radiation falling on a junction causes a current to flow through the external circuit connected to that junction. It is possible to reverse this process and, by making current flow through the junction, obtain radiation from it (Fig. 5.11).

All the mechanisms involved in this INJECTION LUMINESCENCE have already been discussed. When a diode is biased in the forward direction the current is carried by electrons diffusing across the junction from the n-type material to the p-type, and holes diffusing from p to n. Each carrier type finds itself in a dense population of the opposite type and very soon recombines.

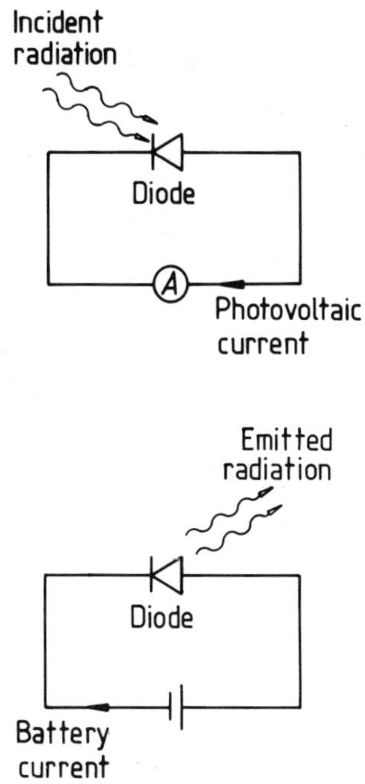

Fig. 5.11 Photovoltaic effect and injection luminescence.

Fig. 5.12 shows an electron A diffusing into the p-type region and about to recombine with a native hole B. Similarly a hole C diffuses towards recombination with an electron D. Now this diagram is a spatial picture of the recombination but we have also to consider the energy-momentum changes that it involves. Referring to Fig. 5.2, each recombination requires an electron at point a to reach a hole at point b. The energy loss the electron suffers in this transition must appear in the form of photons, phonons or both. If the material is direct gap then some will be photon, or radiation, energy; if indirect, mostly phonon, or heat, energy. From this we see that a junction diode made from indirect gap material does not in general form an efficient source of injection luminescence. It is possible, however, to use impurities in an indirect gap material to make a direct or vertical transition possible and so encourage photon production (Fig. 5.13).

The average number of photons produced by one electron-hole recombination (a number always less than one of course) is known as the internal quantum efficiency of a radiation-producing diode. This is greater than the external quantum efficiency, the number of photons which actually leaves the diode, because many may be absorbed by the diode material through which they have to travel.

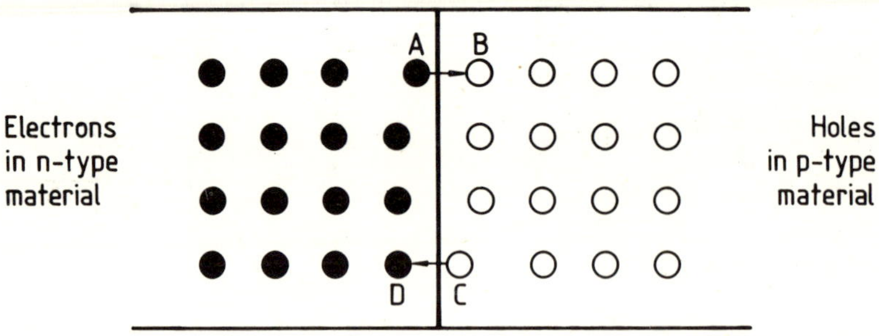

Fig. 5.12 Carrier movements in forward-biased junction.

5.2 THERMAL EFFECTS

5.2.1 Resistance Variation with Temperature

We have seen in Chapter 1 that the resistance of any conducting body depends inversely on the product of the charge carried by the mobile particles, the number of these particles and their mobility. The charge carried by each particle does not change with temperature and may be regarded as a constant for all materials. In a metal the number of carriers is almost independent of temperature and only the small increase in mobility with decreasing temperature causes metal resistivity to be temperature sensitive. The semiconductor, however, shows very large increases in carrier density when the temperature is raised and these dominate the resistivity, over most temperature ranges swamping any small decreases in mobility that the increase in temperature may bring about. The situation is summarised in Table 5.1.

The carrier density, which dominates the resistivity in semiconductors, will depend on its impurity content and its intrinsic properties at various temperatures. Consider a donor-bearing material such as that in Fig. 2.13. At very low temperatures no electrons will be excited from the valence to the conduction band because of the lack

Table 5.1

	Temperature effect on:			
	Electric charge on carrier	Mobility	Number of carriers	Resistivity
Metal	None	Small decrease with temperature increase	None	Small increase with temperature
Semiconductor	None	Small change, either increase or decrease, with temperature	Increase, possibly very rapid, with temperature increase	Large decrease with temperature

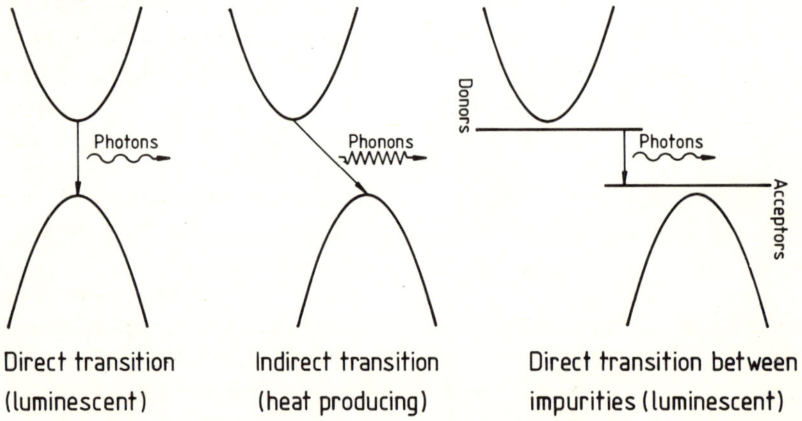

Fig. 5.13 Radiation and heat producing transitions.

of thermal energy. The easier excitation from the donor to the conduction band will dominate and the free carrier population will consist of n electrons given by

$$n \approx (N_D N_C)^{1/2} e^{-E_d/2kT} \qquad \text{(from 2.22)}$$

The resistivity ρ will be:

$$\rho = 1/\mu_n nq = \frac{\mu q}{(N_D N_C)^{1/2}} e^{E_d/2kT} \qquad (5.7a)$$

As the temperature rises the donor centres become exhausted and a region of almost constant resistivity occurs. Eventually, at high temperatures, excitation of electrons from the valence band to the conduction band sets in and the resistivity starts to fall sharply. The density of the free electrons in this region equals the density of the free holes and we have

$$n = p \approx (N_V N_C)^{1/2} e^{-E_g/2kT}$$

and so

$$\rho = 1/(\mu_n n + \mu_p p)q = \frac{(\mu_n + \mu_p)q}{2(N_C N_V)^{1/2}} e^{E_g/2kT} \qquad (5.8a)$$

It is often convenient to work with the natural logarithm of the resistivity and the reciprocal temperature as variables. Equations 5.7a and 5.8a may be rewritten as

$$\log_e \rho = \log_e \frac{\mu q}{(N_D N_C)^{1/2}} + \frac{E_d}{2k}\left(\frac{1}{T}\right) \qquad (5.7b)$$

and

$$\log_e \rho = \log_e \frac{(\mu_n + \mu_p)q}{2(N_C N_V)^{1/2}} + \frac{E_g}{2k}\left(\frac{1}{T}\right) \qquad (5.8b)$$

These considerations are summarised pictorially in Fig. 5.14, wherein $\log_e \rho$ is plotted against $1/T$.

Equations 5.7a and 5.8a show that measurements of resistance as a function of temperature will yield the energy gaps E_D and E_g, or E_A and E_g if a material with dominant acceptor centres is under investigation. The variations in mobility with temperature are too small to interfere with the long regions of linearity in plots of $\log_e \rho$ versus $1/T$. These very large changes of resistance with temperature make semiconductors potentially very useful as thermometers, since electrical resistance is easily measured and displayed by simple circuitry.

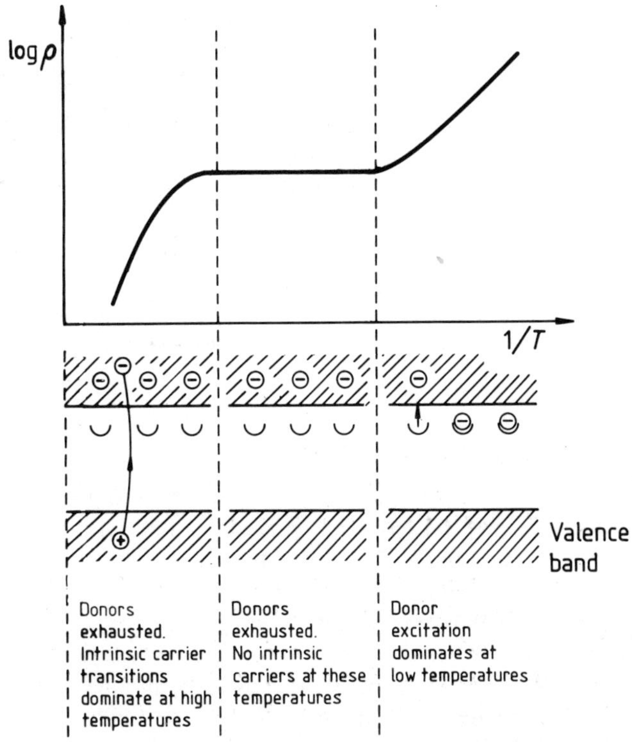

Fig. 5.14 Resistivity as a function of temperature.

5.2.2 The Thermoelectric Effect

If two dissimilar conducting materials are joined in a continuous loop, as shown in Fig. 5.15, and the junctions are maintained at differing temperatures, say T and $T + dT$, an EMF will appear in the circuit. If the EMF is written as $d\varepsilon$, the Seebeck effect or thermoelectric power ϕ is defined by

$$\phi = \frac{d\varepsilon}{dT} \qquad (5.9)$$

A related effect occurs if both junctions are held at a similar temperature T while a small current dI is passed round the circuit. Heat energy will be absorbed at one junction and emitted at the other. If this heating, or cooling, power is dW, then the PELTIER COEFFICIENT π is defined by

$$\pi = \frac{dW}{dI} \qquad (5.10)$$

It must be emphasised that none of the effects considered arises from simple joule heating, but are additional to this.

The thermoelectric power ϕ and the Peltier coefficient π are thermodynamically related by the absolute temperature T so that

$$\phi = \frac{\pi}{T} \qquad (5.11)$$

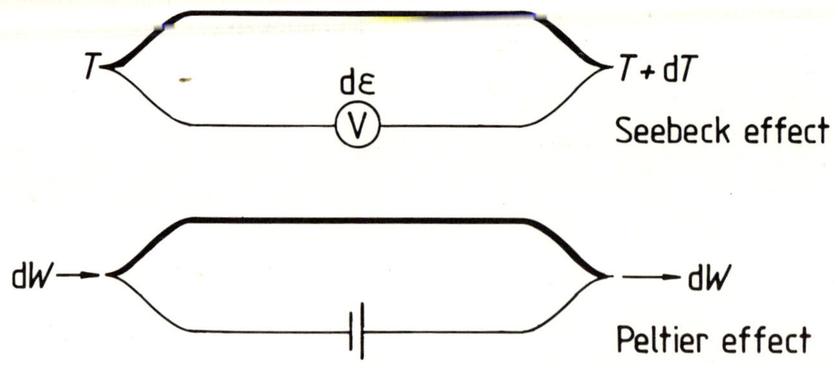

Fig. 5.15 Thermoelectric effects.

This relationship is very useful because π may be quite easily expressed in terms of the fundamental parameters of a metal or semiconductor and the value of ϕ obtained from this. Let the loops in Fig. 5.16 be composed of a metal and an n-type semiconductor. The energy bands in the junction regions will look as drawn in Fig. 5.16, with the fermi level constant (since metal and semiconductor are at the same temperature) and falling in a partly filled band in the metal and in the forbidden energy region in the semiconductor. To obtain a value for π we allow a current to flow across this junction and estimate the power absorbed, or liberated, when it does so.

Let the current have a magnitude such that one electron per second passes the junction. The energy necessary to take an electron from the metal, where it travels near the fermi level, to the semiconductor where it travels at the bottom of the conduction band is E_f eV. So:

$$\mathrm{d}I = q \text{ coulomb/s}$$

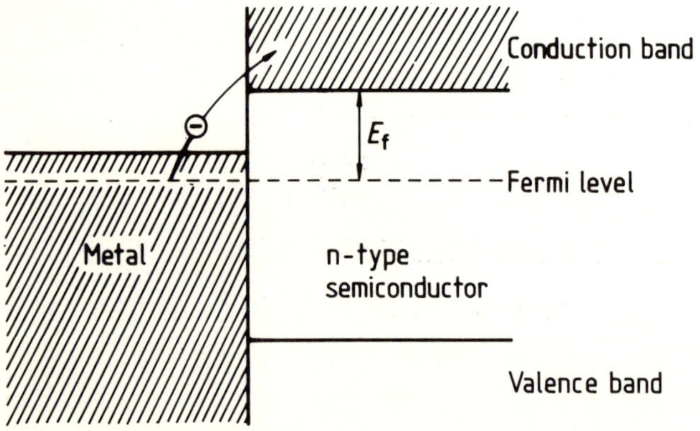

Fig. 5.16 Energy levels involved in thermoelectric effects.

and

$$\mathrm{d}W = -E_f \text{ eV/s}$$

(The negative sign is introduced to conform with the convention that, with an electron current flowing from metal to semiconductor, absorbed energy is negative and emitted energy is of positive sign.)

Whence

$$\pi = \frac{\mathrm{d}W}{\mathrm{d}I} = -\frac{E_f}{q}\,\mathrm{V} \qquad (5.12)$$

The fermi level is related to the carrier density n by

$$n = N_c\,\mathrm{e}^{-E_f/kT} \qquad (2.18a)$$

and so

$$E_f = kT\log_e \frac{N_c}{n}$$

making

$$\pi = -\frac{kT}{q}\log_e \frac{N_c}{n} \qquad (5.13)$$

and

$$\phi = \frac{\pi}{T} = -\frac{k}{q}\log_e \frac{N_c}{n} \qquad (5.14)$$

Had we chosen a p-type semiconductor for the exercise, Equations 5.13 and 5.14 would have been of opposite sign with the fraction N_c/n replaced by N_v/p.

An interesting but rather complex situation obtains when significant quantities of both holes and electrons are present together. With n electrons and p holes having mobilities of μ_n and μ_p respectively, for every electron that passes the junction $\mu_p p/\mu_n n$ holes must pass. By examining Fig. 5.17 it is seen that the energy change necessary to take a hole across the metal-semiconductor

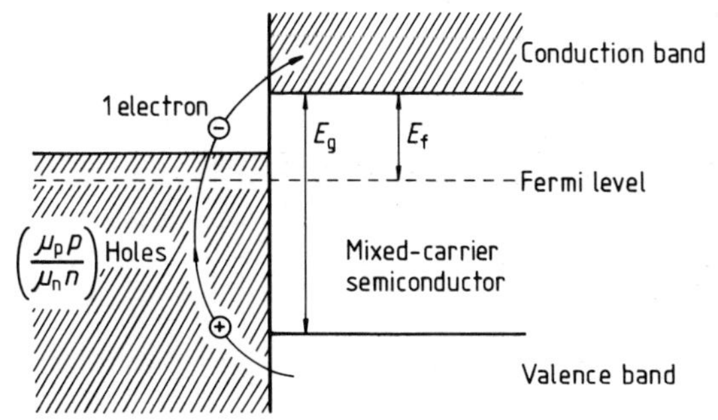

Fig. 5.17 Energy levels involved in mixed-carrier thermoelectric effects.

junction is $+(E_g - E_f)$ eV. Hence in unit time a charge of $q + q(\mu_p p/\mu_n n)$ passes the junction due to hole and electron carriers and dI is given by

$$dI = q\left(1 + \frac{\mu_p p}{\mu_n n}\right)$$

Also in unit time an energy change of $-E_f + (E_g - E_f)(\mu_p p/\mu_n n)$ takes place, giving

$$dW = E_g \frac{\mu_p p}{\mu_n n} - E_f\left(1 + \frac{\mu_p p}{\mu_n n}\right)$$

$$\pi = \frac{dW}{dI} = \frac{E_g \dfrac{\mu_p p}{\mu_n n} - E_f\left(1 + \dfrac{\mu_p p}{\mu_n n}\right)}{q\left(1 + \dfrac{\mu_p p}{\mu_n n}\right)}$$

and

$$\phi = \frac{E_g \dfrac{\mu_p p}{\mu_n n} - E_f\left(1 + \dfrac{\mu_p p}{\mu_n n}\right)}{qT\left(1 + \dfrac{\mu_p p}{\mu_n n}\right)}$$

Substituting from

$$n = N_c e^{-E_f/kT}$$

and

$$p = N_v e^{-(E_g - E_f)/kT}$$

allows ϕ to be written as

$$\phi = -\frac{k}{q}\left(\frac{n\dfrac{\mu_n}{\mu_p}\log\dfrac{N_c}{n} - p\log\dfrac{N_v}{p}}{n\dfrac{\mu_n}{\mu_p} + p}\right) \tag{5.15}$$

It should be confirmed that Equation 5.15 tends to Equation 5.14 when p is small compared with n.

Having examined the three possible metal-semiconductor junction effects it will be easy to write down the thermoelectric power of a p–n junction by examination of Fig. 5.18.

For every electron passing from the n to the p side of the junction $p_p \mu_p/n_n \mu_n$ holes must pass in the opposite direction. Therefore

$$dI = q\left(1 + \frac{\mu_p p_n}{\mu_n n_n}\right)$$

The energy exchange will be

$$-(E_{f+} - E_{f-}) + \frac{\mu_p p_n}{\mu_n n_n}(E_{f+} - E_{f-})$$

and so

$$dW = (E_{f+} - E_{f-})\left(\frac{\mu_p p_p}{\mu_n n_n} - 1\right)$$

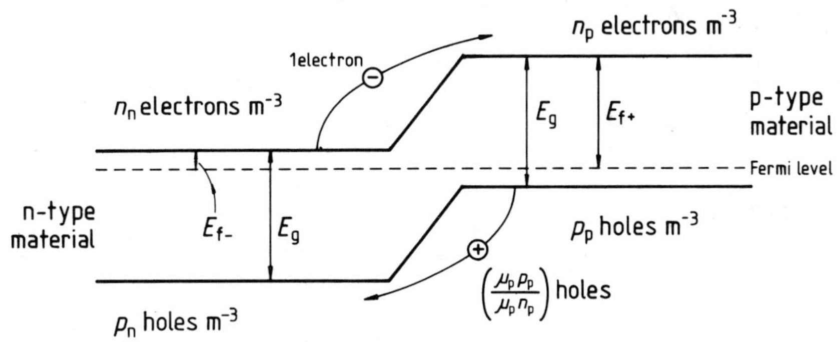

Fig. 5.18 Source of thermoelectric power in p–n junction.

giving

$$\pi = \frac{(E_{f+} - E_{f-})\left(\dfrac{\mu_p p_p}{\mu_n n_n} - 1\right)}{q\left(1 + \dfrac{\mu_p p_p}{\mu_n n_n}\right)}$$

and

$$\phi = \frac{(E_{f+} - E_{f-})\left(\dfrac{\mu_p p_p}{\mu_n n_n} - 1\right)}{qT\left(1 + \dfrac{\mu_p p_p}{\mu_n n_n}\right)}$$

$$= -\frac{k}{q}\left[\log_e\left(\frac{p_p n_n}{n_1^2}\right)\right]\left[\frac{n_n \dfrac{\mu_n}{\mu_p} - p_p}{n_n \dfrac{\mu_n}{\mu_p} + p_p}\right] \quad (5.16)$$

Both the Seebeck and the Peltier effects have found practical use. By measuring the value of ε in the circuit of Fig. 5.15, the temperature difference between the junctions may be known and, if one junction is maintained at a known temperature, the absolute temperature of the other junction can be calculated. Refrigeration can be effected if a current is passed round the circuit so that heat is absorbed from apparatus located near one junction and dissipated at the other junction. Certain photovoltaic detectors are cooled in use to near dry-ice temperatures by this effect.

5.3 MAGNETIC EFFECTS

5.3.1 Introductory: The Lorenz Force

The Hall Effect and Magnetoresistance which will be discussed in this section both arise from the Lorenz force

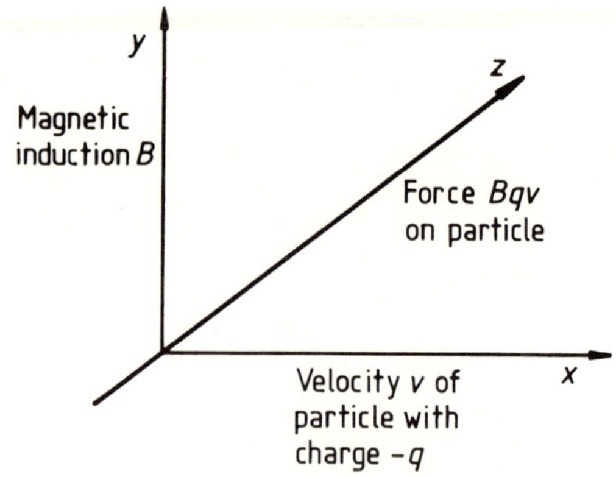

Fig. 5.19 The Lorenz force.

exerted by a magnetic flux on a moving charge. If a hole or electron is moving in the x direction in a system of Cartesian coordinates and a magnetic flux of induction is applied in the y direction, the particle will experience a force acting in the z direction equal to the product of the flux, the particle charge and the particle velocity (Fig. 5.19). It is this force which is used to deflect the electron beam in a cathode ray tube.

5.3.2 The Hall Effect

Consider a uniform slab of n-type semiconductor with the dimensions shown in Fig. 5.20. If a voltage V_L is applied across the end contacts then the carrier velocity v will be $\mu \times$ electric field

$$v = \mu \frac{V_L}{l}$$

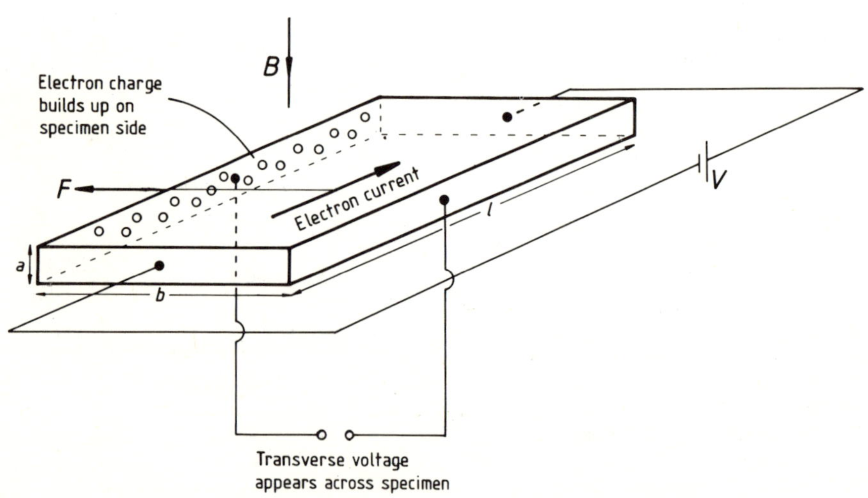

Fig. 5.20 Hall effect in n-type specimen.

By applying the magnetic flux B at right angles to the slab, a force F equal to Bqv is exerted on the carriers

$$F = Bq\mu \frac{V_L}{l}$$

This will drive the electrons across the specimen until they build up a charge which produces a transverse voltage V_T just sufficient to balance the force F.

Transverse field × electron charge = F

$$\frac{V_T q}{b} = Bq\mu \frac{V_L}{l}$$

which can be rearranged to give

$$\mu = \frac{V_T}{V_L} \frac{l}{b} \frac{1}{B} \tag{5.17}$$

All the factors on the right hand side of Equation 5.17 can be measured and so the experiment allows the value of μ to be calculated. The resistivity ρ of the material is easily measured and this allows the carrier density to be obtained using the relationship

$$\frac{1}{\mu n q} = \rho$$

If the specimen contains both holes and electrons it is not possible to set up a transverse electric field that will balance the Lorenz forces on each type of carrier since these will be in the same direction and any transverse electric field that decelerates one carrier will accelerate the other (Fig. 5.21). In these circumstances equilibrium is reached when equal numbers of holes and electrons flow across the specimen, thus carrying a net zero charge.

For the electrons:

Total transverse force = Lorenz force − Transverse field force

$$= Bq\mu_n V_L - \frac{V_T}{b} q$$

Transverse velocity = Transverse force × μ_n/q

$$= \left(Bq\mu_n V_L - \frac{V_T}{b} q\right) \mu_n/q$$

Transverse flux = Transverse velocity × density

$$= \left(Bq\mu_n V_L - \frac{V_T}{b} q\right) \mu_n n/q$$

Similarly for holes

$$\text{Transverse flux} = \left(Bq\mu_p V_L + \frac{V_T}{b} q\right) \mu_p p/q$$

For zero charge transport these two fluxes must be equal:

$$\left(Bq\mu_n V_L - \frac{V_T}{b} q\right) \mu_n n = \left(Bq\mu_p V_L + \frac{V_T}{b} q\right) \mu_p p$$

or

$$\frac{V_T/b}{BV_L} = \frac{n_n{}^2 - p_p{}^2}{n_n + p_p} \tag{5.18}$$

The two expressions for the Hall Effect, Equations 5.17 and 5.18, can both be rewritten in terms of the current I flowing in the specimen. For the n-type material we have

$$V_L = \rho \frac{l}{ab} I$$

$$= \frac{l}{\mu_n n q a b} I$$

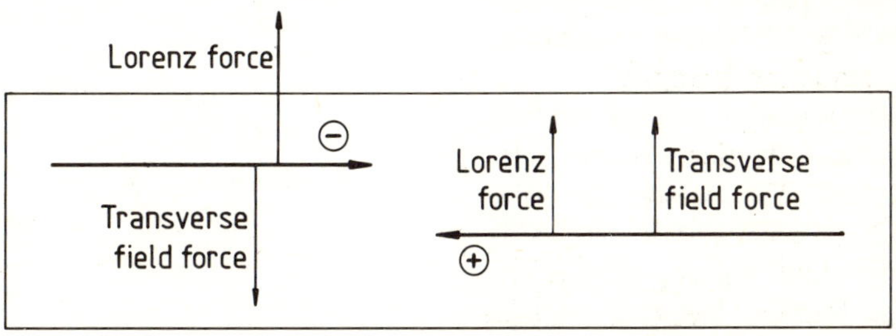

Fig. 5.21 Hall effect forces in mixed-carrier semiconductor.

Putting this into Equation 5.17

$$\mu_n = V_T \frac{\mu_n nqab}{Il} \frac{l}{b} \frac{1}{B}$$

$$\frac{I}{nq} = \frac{V_T}{b} \times \frac{ab}{I} \times \frac{1}{B}$$

If the current density I/ab and the magnetic flux B are both made equal to unity, the measured transverse electric field V_T/b has the magnitude of the HALL COEFFICIENT, R_H

$$\frac{1}{nq} = R_H = \left[\frac{V_T}{b}\right]_{I/ab=B=1} \tag{5.19}$$

The relationship between V_L and I for the mixed conduction case is

$$V_L = \frac{lI}{(n\mu_n + p\mu_p)abq}$$

Putting this into Equation 5.18 gives

$$\frac{n\left(\frac{\mu_n}{\mu_p}\right)^2 - p}{q\left(n\frac{\mu_n}{\mu_p} + p\right)^2} = \frac{V_T}{b} \times \frac{1}{B} + \frac{ab}{I}$$

and as before for unit field and current density

$$\frac{n\left(\frac{\mu_n}{\mu_p}\right)^2 - p}{q\left(n\frac{\mu_n}{\mu_p} + p\right)^2} = R_H = \left[\frac{V_T}{b}\right]_{I/ab=B=1} \tag{5.20}$$

5.3.3 Magnetoresistance

If the resistivity of the slab of semiconductor material discussed in the last section is measured between the two large area end contacts it will be found to increase slightly when the magnetic induction is applied.

Referring to Fig. 5.22 the resistivity ρ_0 measured in the absence of any magnetic induction increases to $\rho_0 + d\rho_0$ when B is applied. The reason for this TRANSVERSE MAGNETORESISTANCE EFFECT lies in the differing mobilities that a group of carriers displays. Although a large fraction has the average mobility $\bar{\mu}$ some carriers have mobilities greater than this and some have mobilities less than $\bar{\mu}$ (Fig. 5.23).

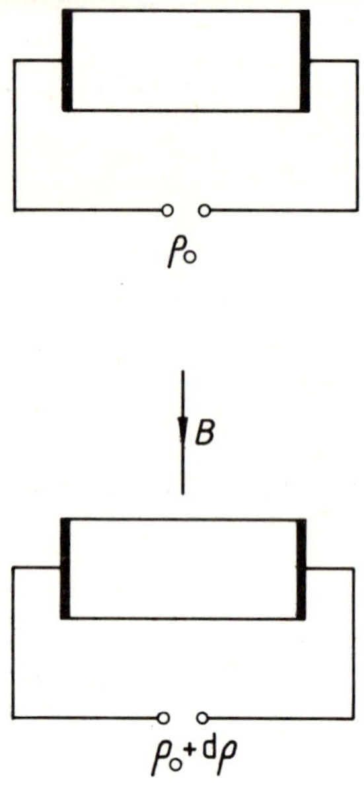

Fig. 5.22 Transverse magnetoresistance.

These carriers will not have their Lorenz force exactly balanced by the transverse Hall field and will travel at an angle to the axis of the specimen thus reducing the effective conductivity or increasing the resistivity (Fig. 5.24).

To simplify analysis we will divide the carriers into three groups:

a small number dn having a mobility $\bar{\mu} - d\mu$,
a small number dn having a mobility $\bar{\mu} + d\mu$, and
$n - 2\,dn$ having the average mobility $\bar{\mu}$.

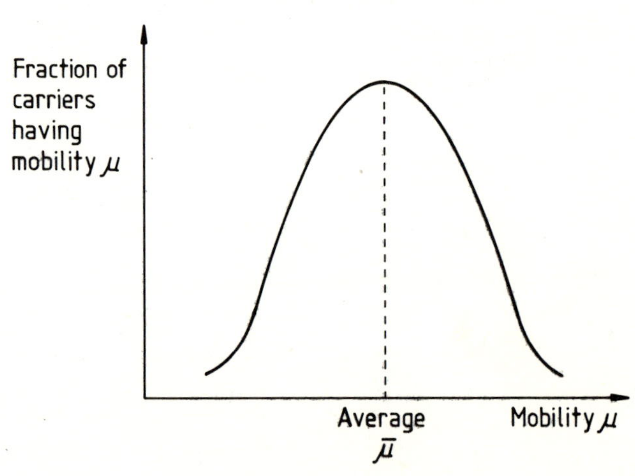

Fig. 5.23 Distribution of mobilities.

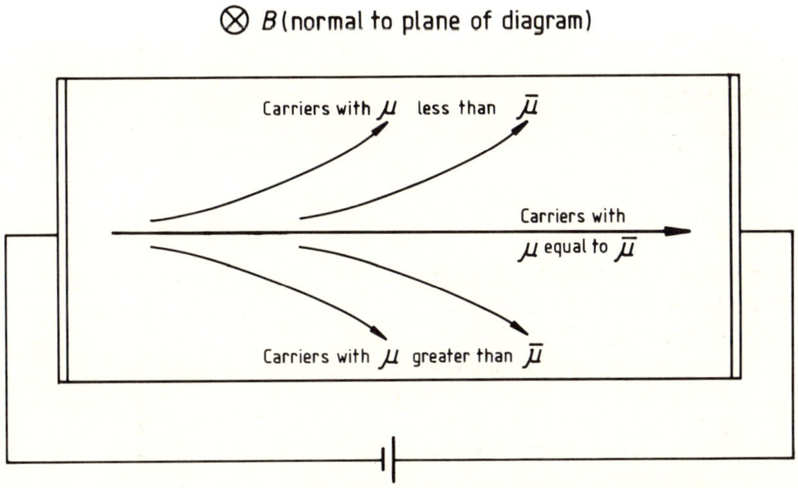

Fig. 5.24 Schematic carrier trajectories for different mobilities.

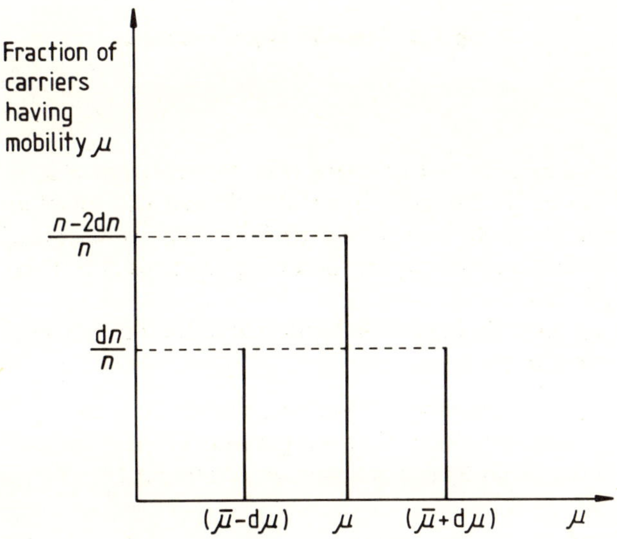

Fig. 5.25 Simplified mobility distribution.

This allows Fig. 5.23 to be redrawn as Fig. 5.25 and Fig. 5.24 to be redrawn as Fig. 5.26.

The total conductivity in the direction of the electric field is made up from three contributions:

$$q[(n - 2\,\mathrm{d}n)\bar{\mu} + \mathrm{d}n(\bar{\mu} - \mathrm{d}\mu)\cos\theta + \mathrm{d}n(\bar{\mu} + \mathrm{d}\mu)\cos\theta]$$
$$= q[\bar{\mu}(n - 2\,\mathrm{d}n + 2\,\mathrm{d}n\cos\theta)]$$

The resistivity $\rho_0 + \mathrm{d}\rho_0$ is thus

$$\rho_0 + \mathrm{d}\rho_0 = 1/q[\bar{\mu}(n - 2\,\mathrm{d}n + 2\,\mathrm{d}n\cos\theta)]$$

The undisturbed resistivity is

$$\rho_0 = 1/q[\bar{\mu}(n - 2\,\mathrm{d}n) + (\bar{\mu} + \mathrm{d}\mu)\,\mathrm{d}n + (\bar{\mu} - \mathrm{d}\mu)\,\mathrm{d}n]$$

From these two equations:

$$\frac{\mathrm{d}\rho}{\rho_0} = \frac{2\,\mathrm{d}n}{n}(1 - \cos\theta) \qquad (5.21)$$

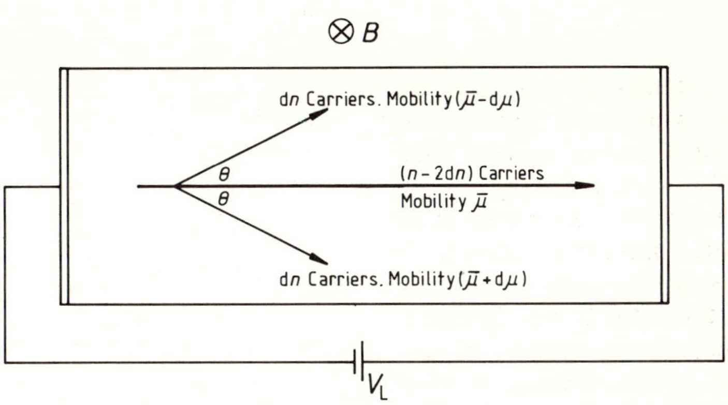

Fig. 5.26 Simplified carrier trajectories for different mobilities.

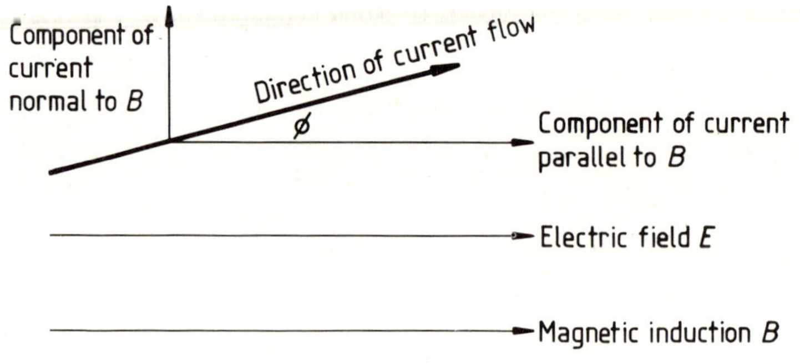

Fig. 5.27 Current flow in material with anisotropic resistivity.

It remains to estimate the value of cos θ. Referring to Fig. 5.26:

$$\text{Total transverse force} = \text{Lorenz force} - \text{Transverse electric force}$$

$$= Bq(\bar{\mu} \pm d\mu)\frac{V_L}{l} - q\frac{V_T}{b}$$

$$= Bq(\bar{\mu} \pm d\mu)\frac{V_L}{l} - Bq\bar{\mu}\frac{V_L}{l}$$

$$= \pm Bq\, d\mu \frac{V_L}{l}$$

$$\text{Longitudinal force} = q\frac{V_L}{l}$$

$$\cos\theta = \frac{\text{Longitudinal force}}{[(\text{Longitudinal force})^2 + (\text{Transverse force})^2]^{1/2}}$$

$$\approx 1 - \tfrac{1}{2}(B\, d\mu)^2$$

Putting this value into Equation 5.21 gives

$$\frac{d\rho}{\rho_0} \approx \frac{B^2}{n} dn(d\mu)^2 \qquad (5.22)$$

The resistivity may also increase if the magnetic field is applied in a direction parallel to the electric field and this LONGITUDINAL MAGNETORESISTANCE EFFECT originates in the anisotropy of the resistivity which many semiconductors show. If the resistivity in the x direction is ρ_x and that in the y direction ρ_y, an electric field applied at some angle α to the x axis will cause a current to flow in a direction which is not parallel to the applied field. Because of this the current will have a velocity component normal to the magnetic field direction and this will be acted on by Lorenz forces (Fig. 5.28).

The angle ϕ between E and the current flow can be obtained by considering Fig. 5.28.

$$\text{Current component } I_x \text{ in } x \text{ direction} = \frac{E\cos\alpha}{\rho_x}$$

$$\text{Current component } I_y \text{ in } y \text{ direction} = \frac{E\sin\alpha}{\rho_y}$$

$$\text{Direction of current flow } \gamma = \tan^{-1}\frac{I_y}{I_x}$$

$$= \tan^{-1}\left(\tan\alpha\frac{\rho_x}{\rho_y}\right)$$

$$\text{Angle } \phi \text{ between } E \text{ and } I = \gamma - \alpha$$

$$\tan\phi = \tan(\gamma - \alpha) = \frac{\tan\alpha - \tan\alpha\dfrac{\rho_x}{\rho_y}}{1 + \tan^2\alpha\dfrac{\rho_x}{\rho_y}} \qquad (5.23)$$

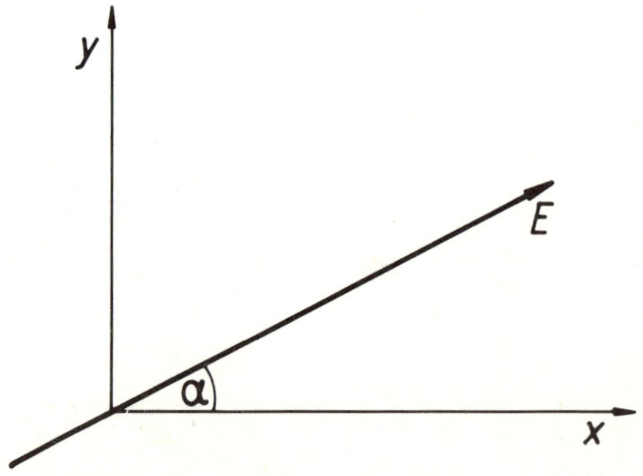

Fig. 5.28 General field direction in anisotropic conductor.

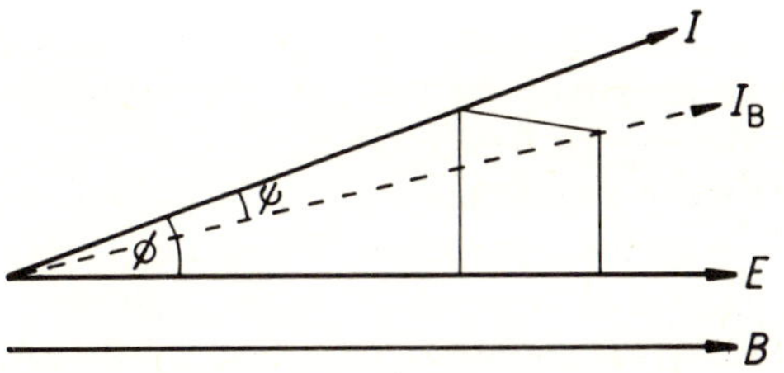

Fig. 5.29 Directions of current flow in magnetoresistive specimen.

We can now attempt a simplified analysis of the effect the magnetic induction B has on the resistivity. Referring to Fig. 5.29, the resolved current density in the direction of E, without magnetic induction, is $I \cos \phi$ therefore

$$\rho_0 = \frac{E}{I \cos \phi}$$

When the magnetic induction B is applied, the direction of current flow swings through some angle ψ due to the Lorenz force on that component of current which is normal to B. The resolved component of the current in the direction of E is now $I \cos \phi \cos \psi$ and

$$\rho_0 + \mathrm{d}\rho = \frac{E}{I \cos \phi \cos \psi}$$

and so

$$\frac{\mathrm{d}\rho}{\rho} = \left(\frac{1}{\cos \psi} - 1\right) \tag{5.24}$$

It remains to find $\cos \psi$. Referring to Fig. 5.30, the particle velocity in the current component $I \sin \phi$ normal to B is $I \sin \phi/qn$ and therefore the Lorenz force is

$$Bq \times \text{velocity} = Bq \frac{I \sin \phi}{qn}$$

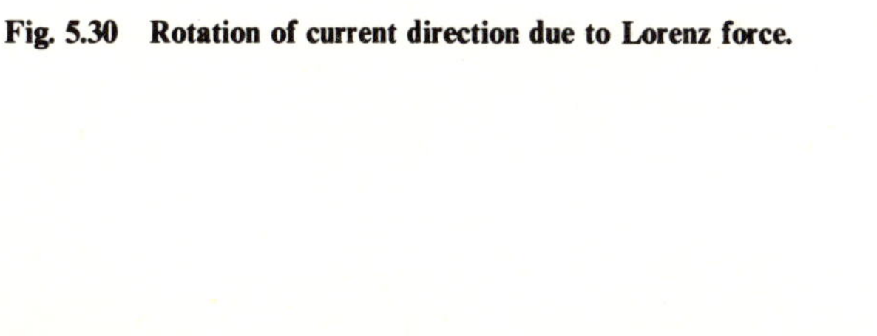

Fig. 5.30 Rotation of current direction due to Lorenz force.

The particle velocity in the undisturbed current $I \cos \phi$, is $I \cos \phi/nq$. Hence

$$\cos \psi = \frac{1}{\sqrt{[1 + (B\mu \tan \phi)^2]}}$$

and

$$\frac{1}{\cos \psi} - 1 \approx \tfrac{1}{2}(B\mu \tan \phi)^2 \tag{5.25}$$

Taking Equation 5.24 and substituting for $(1/\cos \psi) - 1$ from Equation 5.25 and for $\tan \phi$ from Equation 5.23 gives

$$\frac{\mathrm{d}\rho}{\rho_0} = B^2 \frac{(\mu \tan \alpha)^2}{2} \left(\frac{1 - \dfrac{\rho_x}{\rho_y}}{1 + \tan^2 \alpha \dfrac{\rho_x}{\rho_y}} \right) \tag{5.26}$$

While Equations 5.22 and 5.26 emphasise the factors that give rise to the two magnetoresistance effects, they are greatly simplified compared with the results that rigorous analysis yields. The precise relationships allow experimental results to be interpreted in terms of the variation of effective mass of the carriers with their direction of travel in the semiconductor crystal.

PROBLEMS

For values of required constants, see Problems to Chapter 1.

5.1 Radiation falling on one plane face of a semiconducting disc generates free electrons at a rate of 10^{19} m^{-2} s^{-1}. If this increases the conductivity of the disc by 0.10%, what is the lifetime of a free electron? The disc is 10^{-3} m thick and has a carrier population of 10^{22} electrons m^{-3} in the dark.

5.2 A photosensitive diode operated at 300 °K shows a short circuit current of 10^{-3} A when light falls upon it. If the open circuit voltage is 0.5 V, what is I_0, the saturation current of the diode?

5.3 A current of 10^{-3} A is passed through a light-emitting diode, causing the emission of radiation with a wavelength of 1.24 microns. If the efficiency of the device is 30%, what is the power of the emitted radiation?

5.4 A specimen of intrinsic semiconductor has a resistance of 1 Ω at 77 °K, falling to 0.0308 Ω at 100 °K. What is the approximate energy gap of the material?

5.5 The electron mobility is twice the hole mobility in an intrinsic semiconductor. If the thermoelectric power is -5.56×10^{-4} V °K^{-1} at 300 °K, what is the energy gap of the material?

5.6 An electric current is carried down a long thin bar of square cross-section by electrons having a mobility of 0.3 m^2 V^{-1} s^{-1}. If a magnetic flux of induction of 0.75 T is applied at right angles to one of the large area faces of the bar, through what angle will the electric field rotate?

5.7 The resistivity of a semiconductor in the x direction is twice that in the y direction. When the electric field is applied at an angle of 10° to the x direction, the longitudinal magnetoresistance effect is 0.010. At what angle must the electric field be applied to increase this value to 0.015 if the magnetic flux of induction is kept constant?

CHAPTER 6

Unipolar Devices

When large electric fields are applied to the surface of a semiconductor, with an insulating layer placed to prevent a continuous current flow, electrons will be either attracted towards, or repelled from, the surface, according to the direction of the field. The attraction or repulsion varies the local resistivity of the semiconductor by changing the number of free electrons in it. The intrusion of a junction depletion region into a layer, or slice, of semiconductor may also be used to change its resistance.

This chapter describes how these mechanisms are made use of in the Field Effect Transistor, a device which is gradually replacing the conventional transistor described in Chapter 4.

6.1 MAJORITY CARRIER UNIPOLAR DEVICES

6.1.1 Introductory

In this chapter we must avoid becoming too alarmed by the polysyllabic titles and mysterious acronyms which modern electronic devices assume. A unipolar device is simply one in which only one type of carrier (hole or electron) is active, as opposed to the bipolar transistors already discussed in which both carrier types function. In this section we shall consider the class of unipolar transistors in which the majority carriers, holes in p-type and electrons in n-type material, are effective.

Field Effect Transistors (FETs) are essentially unipolar and those we now consider are also majority carrier devices. They operate as series resistors with 'blocking' layers which are controlled by the magnitude of the signal applied to the transistor. The power needed to vary these blocking layers is very small and can be used to control large currents passing through the device and so provide amplification (Fig. 6.1). Although the device was proposed by Lilienfeld and others in 1925, industry has only recently begun to exploit it. They are more similar in concept to the thermionic valve than bipolar transistors, and, indeeed, their characteristics as circuit elements closely follow those of a pentode valve.

6.1.2 The Junction FET

Consider the bar of n-type semiconductor in Fig. 6.2 with a p-type block in the centre. This will be surrounded by a high-resistance p-n junction region

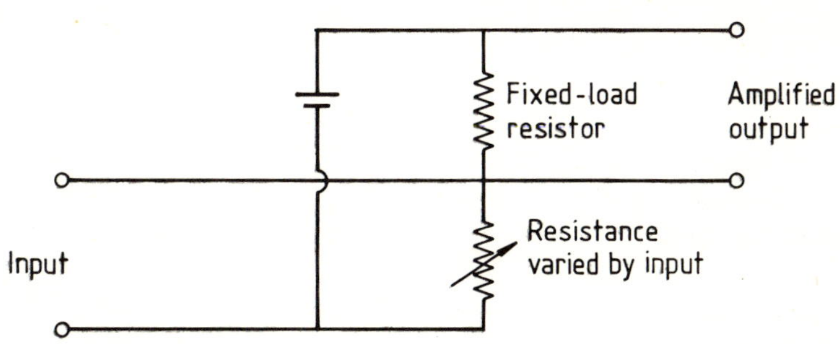

Fig. 6.1 Amplification by variable resistor.

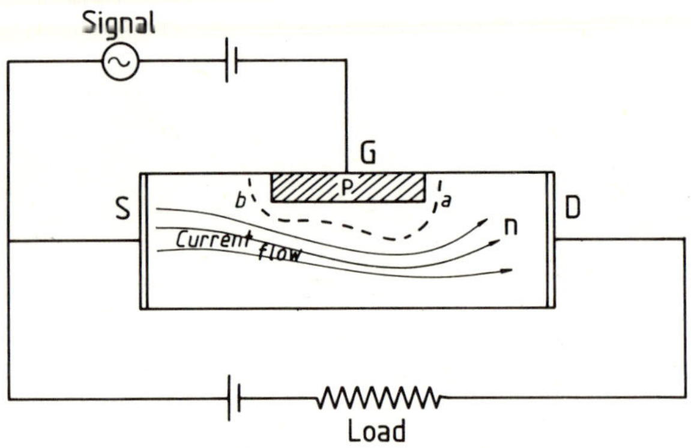

Fig. 6.2 Junction FET.

which will effectively insulate it from the n-type material. Current flowing through the bar between the end contacts S and D will have to pass through the narrow region in the centre. If a reverse bias is now applied to the junction as shown, the width will increase and so reduce the current path and increase the resistance between the end contacts. Conventionally, the negative terminal S is called the SOURCE and the positive terminal D is called the DRAIN while the junction electrode G is termed the GATE. These terms emphasise the mechanism of the transistor with electrons originating at the source and passing through a variable gate to reach the drain electrode. Because of the field that is imposed along the bar by the source-to-drain voltage the right hand side of the junction at *a* will always be at a more positive potential than the left hand side at *b*. When the reverse bias is applied to the gate the potential drop across the junction will be greatest therefore at *a* and decrease towards *b*. The junction edge thus reaches out into the n-type bar in the roughly wedge shaped form shown.

It will be seen from this that the effective resistance of the sink-to-drain circuit will increase if either the negative bias on the gate is increased or the sink-to-drain voltage is increased, since both of these actions cause the junction edge to project further into the small central conducting channel.

We can make a simple analysis of this effect. Referring to Fig. 6.3:

ρ = resistivity of the n-type material
t = distance between unbiassed junction edge and the edge of the channel
W_x = width of junction at the point x
a = length of junction
V_g = potential of gate electrode
V_x = potential at x referred to V_g as 0.
h = depth normal to the plane of the diagram

The voltage drop dV_x along a length dx of the channel is given by

$$dV_x = I_{SD}\, dR_x \tag{6.1}$$

where I_{SD} is the channel current and dR_x the resistance of a length dx of channel.
Now the junction width at x is given by

$$W_x = \sqrt{\left[\frac{2\varepsilon}{N_d q}(V_0 + V_g + V_x)\right]} \tag{6.2}$$

where the equation is that for an abrupt junction with N_a, the doping in the p-type gate, much greater than N_d the donor density in the body of the device.

$$dR_x = \frac{\rho\, dx}{h(t - W_x)} = \frac{\rho\, dx}{h\left\{t - \sqrt{\left[\frac{2\varepsilon}{N_d q}(V_0 + V_g + V_x)\right]}\right\}} \tag{6.3}$$

From 6.1 and 6.3

$$dV_x = I_{SD}\frac{\rho\, dx}{h\left\{t - \sqrt{\left[\frac{2\varepsilon}{N_d q}(V_0 + V_g + V_x)\right]}\right\}}$$

Integrating along the length of the channel

$$\int_0^{V_{SD}} h\left\{t - \sqrt{\left[\frac{2\varepsilon}{N_d q}\right](V_0 + V_g + V_x)}\right\} dV_x = \int_0^a I_{SD}\,\rho\, dx$$

giving

$$I_{SD} = V_{SD}\, h\, \frac{t - \sqrt{\left(\frac{8\varepsilon V_{SD}}{9N_d q}\right)}\left\{\left(1 + \frac{V_0 + V_g}{V_{SD}}\right)^{3/2} - \left(\frac{V_0 + V_g}{V_{SD}}\right)^{3/2}\right\}}{\rho a}$$

or

$$R_C = \frac{V_{SD}}{I_{SD}}$$

$$= \frac{\rho a}{h\left[t - \sqrt{\left(\frac{8\varepsilon V_{SD}}{9N_d q}\right)}\left\{\left(1 + \frac{V_0 + V_g}{V_{SD}}\right)^{3/2} - \left(\frac{V_0 + V_g}{V_{SD}}\right)^{3/2}\right\}\right]} \tag{6.4}$$

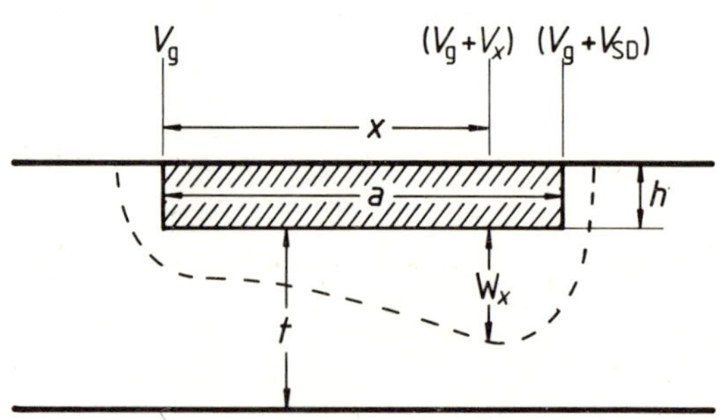

Fig. 6.3 Junction region in FET.

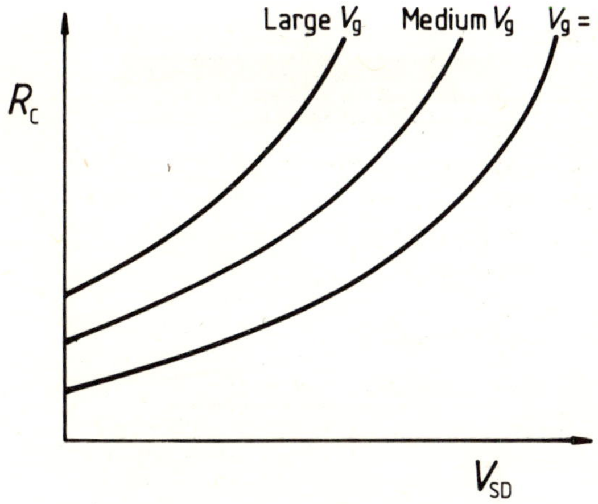

Fig. 6.4 Junction FET resistance.

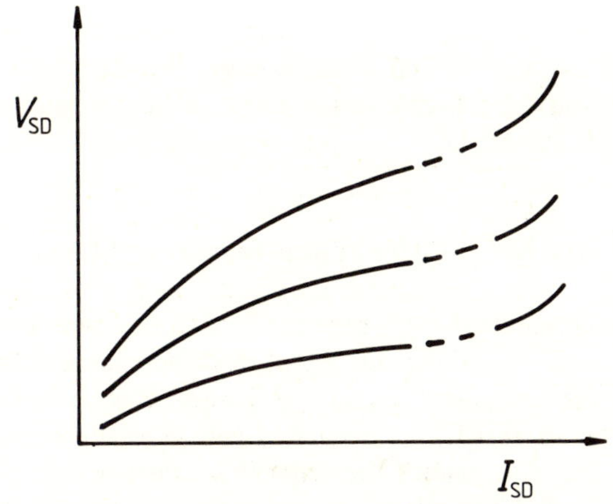

Fig. 6.5 Junction FET current/voltage curves.

The shape of the channel resistance R_C as a function of the source-to-drain voltage V_{SD} is shown in Fig. 6.4 with the gate voltage V_g as a parameter.

The onset of the very large R_C values at high V_{SD} occurs when the junction edge completely blocks or PINCHES OFF the channel and this prevents the current rising beyond a saturation value. It is conventional to display the information in the curves of Fig. 6.4 in the form of plots of the source-to-drain current I_{SD}. This is done in Fig. 6.5 which may be compared with the anode-voltage/anode-current curves of a thermionic valve with the grid-bias as a parameter.

The p–n junction in this device can be replaced by a Schottky junction as shown in Fig. 6.6. The p-type layer gives mechanical strength to the thin n-layer, the junction between the two materials effectively preventing any current flow in that part of the device. Both forms of the FET are used for amplification and currently made from silicon or gallium arsenide. As the latter material can be made high-resistance or semi-insulating it is not necessary in this case to make an isolating junction between the active part of the device and its mechanical support.

6.2 MINORITY CARRIER UNIPOLAR DEVICES

6.2.1 Introductory

In minority carrier devices a locally high concentration of minority carriers is induced in the material by the application of a large electrostatic field. The density of minority carriers produced by the field is large enough to change the material type, say from p to n, and to destroy an existing p–n junction. The removal of the

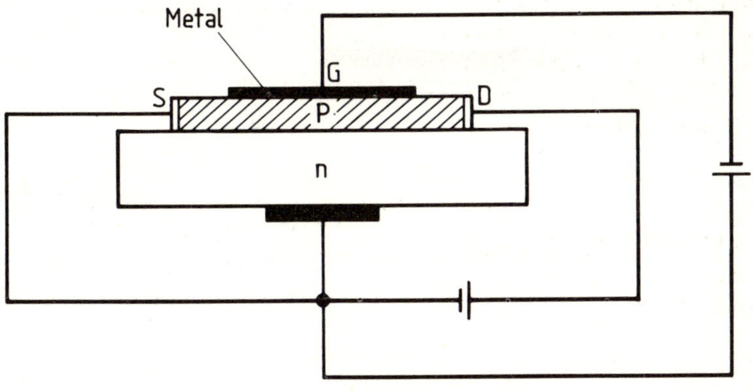

Fig. 6.6 Schottky junction FET.

junction changes a high resistance path to a low resistance and so the small energy required for the electrostatic field can cause large current variations through the device.

6.2.2 The Metal-Oxide-Silicon-Transistor (MOST)

Our initial warning is relevant here as this can be met with as a MOSFET or metal-oxide-semiconducting-field-effect-transistor or, in the United States, it may become an IGFET or insulated-gate-field-effect-transistor. Fig. 6.7a shows the schematic construction of a MOST. A layer of n-type material has two p-type regions set in it to form the source and drain electrodes S and D. An insulating layer of silicon oxide covers the top of the device and separates the metal gate electrode G from the semiconductor material while a large ohmic contact C covers the bottom of the device. With no potential on the gate the source and drain are insulated from one another by the p–n junctions which surround them.

When a negative potential is applied to G the few holes in the n-type slab are drawn up to the interface between the insulating layer and the semiconductor, to form a thin p-type channel (Fig. 6.7b). This channel now forms a conducting link between S and D allowing current to flow in the external circuit. The channel resistance R_C can be obtained by considering the charge density which the gate potential V_G induces in the conducting channel (Fig. 6.8).

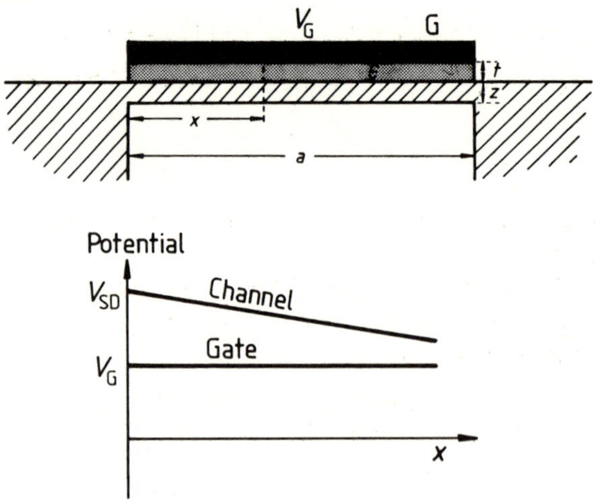

Fig. 6.8 MOST analysis.

Let the insulating layer have a thickness t, a permittivity ε_i, and a depth h. The electrode G and the p-type semiconductor then form a capacitor of capacity C per unit area where

$$C = \frac{\varepsilon_i}{t}$$

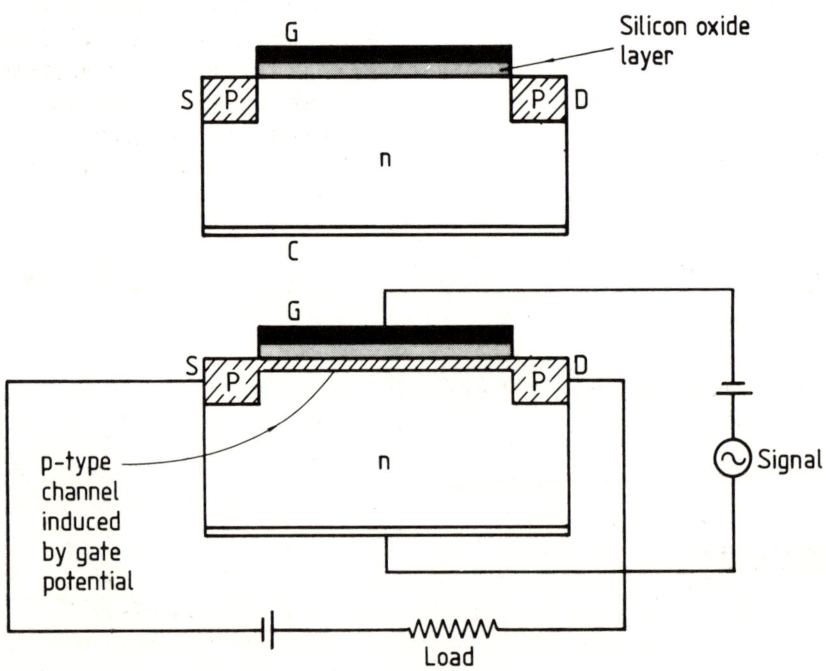

Fig. 6.7 MOST (enhancement mode).

The charge per unit area δ_x, at x in the channel, is then

$$\delta_x = \text{capacity per unit area } x \text{ potential difference}$$

$$= \frac{\varepsilon_i}{t} \times (V_x - V_g)$$

where V_x is the potential at x due to the source-drain voltage.

If the channel containing the charge has a thickness z, then σ_x the charge density in an element dx of the channel is given by

$$\sigma_x = \frac{\text{Charge/unit area} \times \text{Area}}{\text{Volume}} = \frac{\delta_x \cdot h\,\mathrm{d}x}{Zh\,\mathrm{d}x}$$

$$= \frac{\varepsilon_i}{zt}(V_x - V_g)$$

The resistance dR_x of the element is given by

$$\mathrm{d}R_x = \frac{\mathrm{d}x}{\sigma_x zh\mu} = \frac{t\,\mathrm{d}x}{\varepsilon_i(V_x - V_g)\mu h}$$

But

$$\mathrm{d}R_x = \frac{\mathrm{d}V_x}{I_{SD}}$$

Therefore

$$\frac{\mathrm{d}V_x}{I_{SD}} = \frac{t\,\mathrm{d}x}{(V_x - V_g)\mu h\varepsilon_i}$$

Rearranging and integrating:

$$\int_0^{V_{SD}} \varepsilon_i(V_x - V_g)\mu h\,\mathrm{d}V_x = \int_0^a I_{SD}\,t\,\mathrm{d}x$$

$$\varepsilon_i\left(\frac{V_{SD}{}^2}{2} - V_g V_{SD}\right)\mu h = I_{SD}\,ta$$

and so

$$R_C = \frac{V_{SD}}{I_{SD}} = \frac{at}{\varepsilon_i\,\mu h\left(\frac{V_{SD}}{2} - V_g\right)} \qquad (6.6)$$

As before the apparently infinite resistance obtained when $V_{SD} = 2V_g$ is not obtained in practice.

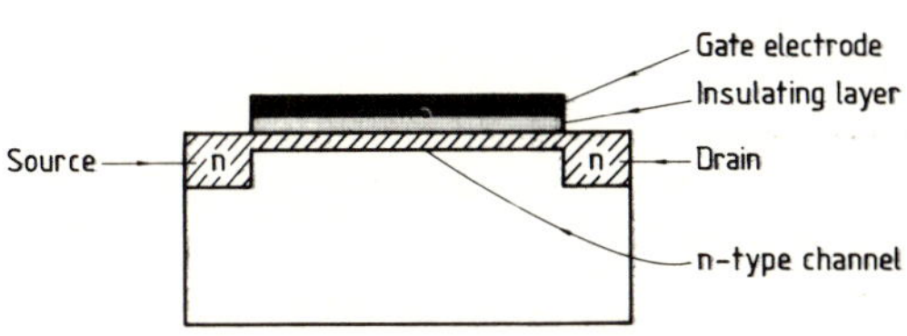

Fig. 6.9 Depletion mode MOST.

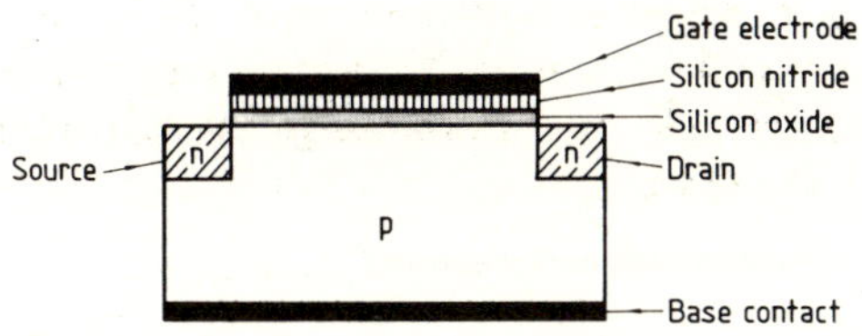

Fig. 6.10 MNOS.

This device just described is known as an ENHANCEMENT MODE MOST as the conductance in the channel is enhanced by the application of the gate voltage. A more efficient form, the DEPLETION MODE, has an n-type conducting channel built into it as shown in Fig. 6.9. This channel is then depleted of its electrons by applying a negative potential to the gate electrode and so reducing the channel conductance. Both forms of MOST are presently constructed in silicon and used as switches or amplifiers.

6.2.3 The Metal-Nitride-Oxide-Silicon Switch (MNOS)

The MNOS, sketched in Fig. 6.10, may be regarded in construction as an enhancement mode MOST with an additional layer of silicon nitride between the gate electrode and the silicon oxide layer.

With moderate voltages applied to the gate the source-drain current I_{SD} varies with the gate voltage V_g in the normal manner as shown in curve a, Fig. 6.11. If now a pulse of high positive voltage is applied to the gate, electrons tunnel through the oxide layer and are trapped in centres in the nitride layer. These remain in position

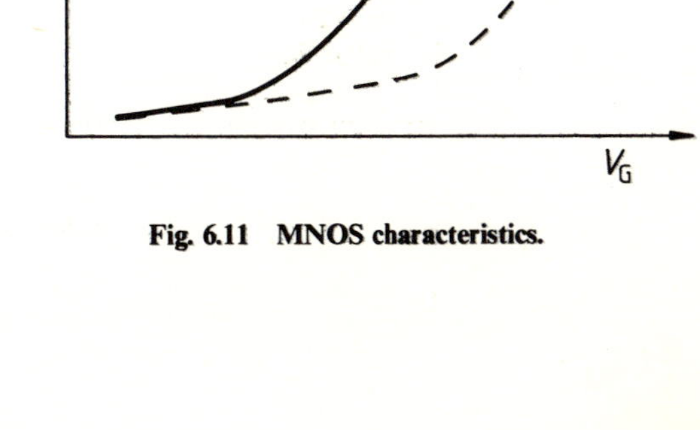

Fig. 6.11 MNOS characteristics.

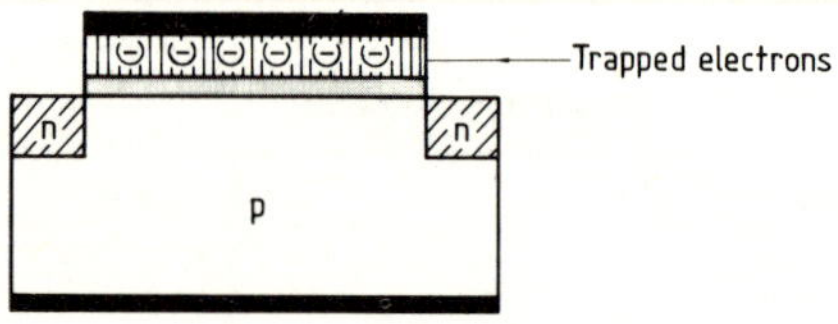

Fig. 6.12 MNOS in activated condition.

indefinitely and by their presence repel other electrons from the oxide-semiconductor interface (Fig. 6.12). Because of this it is difficult to form the n-type channel between source and drain and the I_{SD}/V_g curve shifts to position *b* in Fig. 6.11.

The MNOS is presently constructed in silicon and used as a circuit memory element since it 'remembers' whether a high-voltage pulse has been applied and this information can be elicited (or the device can be 'interrogated') by observing which form of I_{SD}/V_g curve obtains. The initial state of the device can be regained if the traps are discharged by a reverse-voltage pulse.

PROBLEMS

6.1 Express in symbols the source-drain resistance of an FET when the source-drain voltage tends to zero. Show that Equations 6.3 and 6.4 both lead to this expression.

6.2 Express in symbols the differential source-drain resistance of an enhancement mode MOST.

CHAPTER 7

Microwave Devices

Ordinary radio waves can be generated by the conventional use of valves or transistors in circuits containing capacitance and inductance. The electron streams in the circuits are reversed in direction between 100 000 and ten million times a second, causing a radiated electromagnetic wave of between 30 and 3000 metres in length.

Radiation with a wavelength of only a few centimetres has certain advantages over the longer radio wave. These Microwaves, as they are called, can be passed with very little loss along metal tubes, focussed by mirrors or lenses and generally manipulated very much as visible light may be. It is those properties that allow them to be used as the 'searchlight' in a radar scanning system.

It is not possible to generate microwaves by reversing electron streams, as in the generation of radio waves, since the reversal would have to be carried out 10 000 million times a second. Even a particle as light as an electron cannot be conveniently swung around as rapidly as this and it has been necessary to evolve an entirely new technique for producing microwaves.

In essence, all microwave generation involves the leading of bunches of electrons at a high velocity past sets of electrodes, each bunch momentarily inducing a charge on the electrode it is passing. A circuit connected to the electrodes is adjusted to oscillate in response to these repeated pulses of induced charge.

This chapter deals with the manner in which electrons are constrained to form into bunches and the bunches then forced to travel rapidly along an electrode structure.

7.1 MICROWAVES IN THE ELECTROMAGNETIC SPECTRUM

7.1.1 DC to X-rays

The spectrum of electromagnetic energy embraces frequencies from zero to more than 10^{16} Hz, having modes of generation, transmission and interaction with matter which differ not only in scale but in fundamentals over this wide range.

Zero frequency or d.c. energy generation may be by chemical means in batteries or by simple dynamos. Transmission is customarily along conducting rods or wires to some form of resistive load where the energy is usefully transformed to heat, light or mechanical movement.

Electromagnetic energy at frequencies up to 10^2 Hz is obtained from conductors in rotating machinery when they move relative to a magnetic field and so subject the free carriers within to Lorenz forces. These drive the carriers through an external circuit, again using wires for transmission to some form of resistive and reactive load. Domestic and industrial power is generated with frequencies of this order.

Mechanical inertia limits machinery rotation to a few hundred revolutions per second and so, for frequencies between 10^3 and 10^5 Hz it is necessary to cause free, or mobile, electrons to oscillate in semiconducting crystals (transistor oscillators) or in vacuum (thermionic valve oscillators).

Transmission of the generated energy by the simple bare wires already considered would be very inefficient since such wires would radiate energy rather as a wireless transmitting aerial does. To confine this leaking energy the conductors are made in the form of a central rod surrounded by a cylindrical sheath. In such a coaxial cable the current reversals in rod and sheath take place in opposite directions and tend to cancel any radiated field. The generated energy in this frequency range is usually small and used to convey information rather than power.

Much the same applies to generation and local transmission between 10^5 and 10^9 Hz but now if the leaking radiation is encouraged rather than confined it will travel through free space for substantial distances. The radiation-encouraging part of the circuit, basically a bare wire set high above the ground, is known as an aerial and the whole forms the familiar wireless—telephony and television communication system. Just as an oscillating current in an elevated wire radiates electromagnetic energy, so the inverse effect occurs and electromagnetic energy incident on such a wire will induce an oscillating current in it. Thus a second aerial can be used for the remote collection of some fraction of the radiated energy.

From 10^{13} to 10^{17} Hz we pass through the region of infrared, visible light, ultraviolet and finally to X-rays. Generation in this region is not possible by causing electrons to oscillate either in conductors or vacuum, much less by mechanical movement, for the accelerations required would be too great. Instead the movement of electrons between various energy levels within the atom is encouraged. When these energy transitions take place in the orbits of the atom they give rise, as we have seen, to packets of light energy or photons. If they occur within the atomic nucleus the 'packet' is much more energetic and becomes an X-ray. Energy in this frequency range also travels through free space or vacuum without the use of conductors, although it can be guided by reflecting surfaces.

7.1.2 Microwaves

It will have been noted that we have avoided discussing the 10^{10}–10^{11} Hz region in the previous section. This very interesting border land between high-frequency radio waves and long infrared radiation forms the MICROWAVE spectrum with wavelengths between a few millimetres and a few centimetres. While atomic transitions will give too energetic a radiation, that is one with too short a wavelength, the rate of acceleration that would be required of electrons in an oscillating thermionic valve or transistor is precluded by the inertia of the electron. It is possible to make vacuum devices that give these frequencies by causing a series of electron bunches to pass along a row of electrodes (Fig. 7.1). As each bunch passes between a pair of electrodes a current pulse is induced in the external circuit and the repetition frequency, f, of these pulses is the ratio of the bunch velocity V to the inter-pulse separation l:

$$f = \frac{V}{l}$$

This technique of avoiding acceleration of the electron flight is the basis of most vacuum microwave generators. The electrodes may be in a line for the KLYSTRON

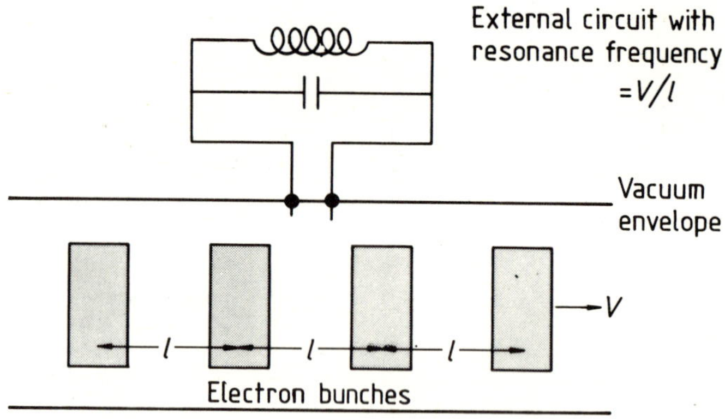

Fig. 7.1 Principle of vacuum microwave generator.

or set in a circle for the MAGNETRON. In the latter a magnetic field normal to the plane of the electron path constrains it to a circular flight by Lorenz forces. The external circuits are usually made to be resonant at the pulse frequency V/l.

The transmission of microwave energy can be by means of WAVEGUIDES. These are conducting tubes of round or rectangular cross-section having diameters or widths of the same order as the radiation wavelength. The microwaves pass along the tube rather as light would pass down a very efficient light-pipe and, if the waveguide in interrupted, energy will be radiated from the open end and can then be directed and focussed by metal mirrors. Essentially then, microwaves are generated by electronic techniques and then manipulated rather as light energy would be.

7.2 SOLID-STATE MICROWAVE GENERATORS

7.2.1 The Gunn Diode

The Gunn diode is analogous to the vacuum microwave generators in that it forms electron bunches, now known as domains, which pass between two electrodes to give the time-varying current required. The interest in this device lies in the manner in which the electron bunches are formed in a semiconductor, usually gallium arsenide.

We will redraw the conduction band from Fig. 5.2 to emphasise the significant aspects for our analysis (Fig. 7.2).

Imagine an electron at a. If a strong electric field is applied, the electron will gain energy until eventually it

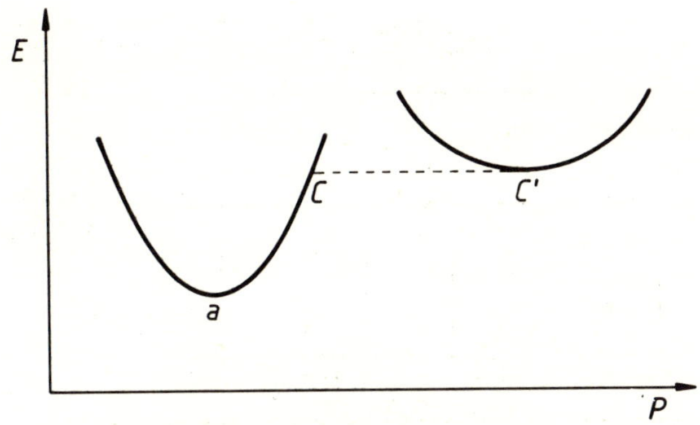

Fig. 7.2 Conduction bands in GaAs.

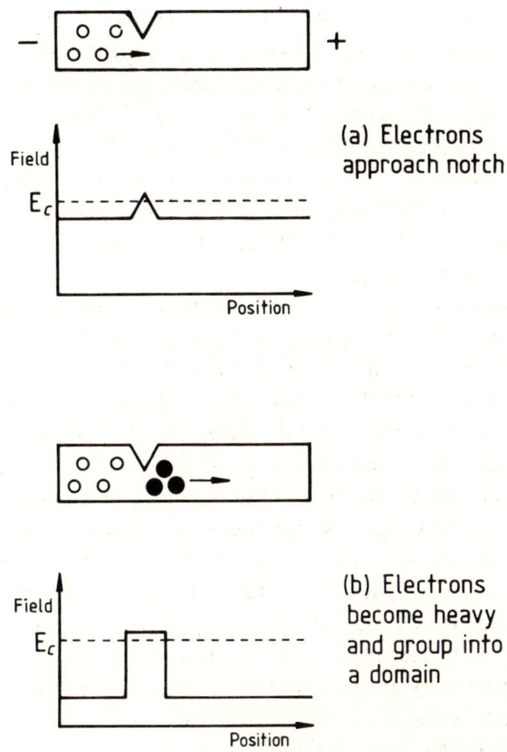

Fig. 7.4 Gunn effect.

reaches the point C. The electron can now transfer to the point C' on another part of the band without any change of energy, although a momentum increase is necessary.

Now at any finite temperature there is momentum available from the vibrations of the heavy lattice atoms and so the transition C to C' is quite probable. When the electron reaches C′ it is in a region where the E/p curvature is larger than at C and hence the effective mass is greater. In short, then, when a field greater than some critical value E_C is applied the electron becomes heavier and hence travels more slowly in the field (Fig. 7.3).

We will now see how this effect is used to produce domains of heavy electrons in a bar of gallium arsenide. The bar is provided with a notch part of the way along and ohmic contacts at each end (Fig. 7.4a).

A voltage is applied across the contacts sufficient to produce a field E just less than E_C in the body of the device but giving a small region in the notch where E is greater than E_C. Electrons drifting from the negative electrode will enter the notch where they experience the higher field and become heavy. They will start to build up a 'plug' of slow, high-resistance material (Fig. 7.4b). Because of its high-resistance most of the potential drop in the bar will appear across it and thus no further production of heavy electrons will be possible. The contact-to-contact resistance of the bar is now great and only a small current flows, a state of affairs that continues until the domain has drifted onto the positive electrodes where it will discharge. The current will now rise until the next domain begins to form, the shape of the current-time plot being as shown in Fig. 7.5. This current is passed into a resonant circuit which is thus triggered into oscillation at the frequency $1/\tau$ where

$$\tau = \frac{\text{Device length}}{\text{Domain velocity}}$$

$$= \frac{\text{Device length}}{\text{`Heavy' electron mobility} \times \text{Electric field in device}}$$

The magnitude of the electric field is of course the value inside the domain.

In practice it is not necessary to introduce an artificial notch as the semiconductor properties are sufficiently disturbed by the presence of the anode contact to give a local high field region there. A typical device would

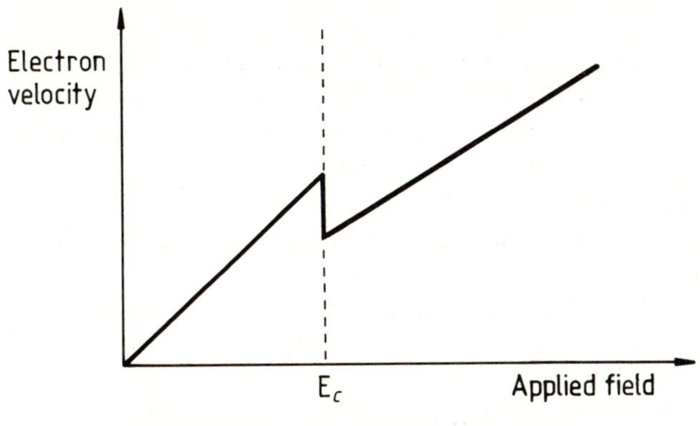

Fig. 7.3 Electron velocities in GaAs.

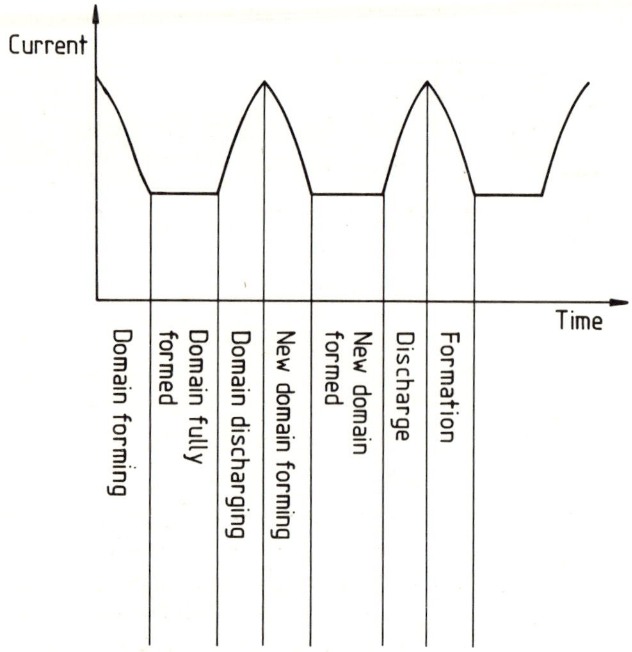

Fig. 7.5 Current in Gunn oscillator.

have a length of 10^{-5} m, an electron mobility of 10 m^2/Vs and a potential difference of 10 V to make $\tau = 10^{-10}$ S and so give a frequency of 10^{10} Hz. Gunn diodes are made commercially from GaAs and experimentally in InP.

7.2.2 IMPATT Avalanche Oscillator

The IMPATT (Impact Avalanche Transit Time) mechanism is conveniently split into two parts.

First consider a diode with a steady reverse bias applied just sufficient to produce the avalanche field within the device. A further, small, a.c. field is now applied so that the total field is swung first above and then below the avalanching value (Fig. 7.6). Each time the total field rises above the avalanche value a bunch of electrons is produced by the impact processes previously described.

The second part of the mechanism consists of a region of semiconductor in which a field slightly above the saturation value is applied. It will be recalled that the saturation field is that which accelerates the carriers to their maximum possible speed.

The bunches of electrons are introduced to this saturation field region and drift across at a steady velocity unaffected by small changes of field. The dimensions of the device are such that each bunch traverses the second region in a time equal to half the period of the small a.c. field but exactly out of phase with it (Fig. 7.7). The bursts of current through the device coincide with the decrease of a.c. field applied.

Now the power absorbed by the device is the product of the current and voltage. The current, however, flows only when the a.c. voltage is in a direction opposite to that of the current flow, therefore the a.c. power absorbed is negative, i.e. power is generated.

In practice the two separate devices we have considered are made as adjacent regions in a single diode, the bunches of avalanche electrons being pumped straight into the saturation field region. The need to apply an a.c. voltage to what is meant to be an oscllator may seem to show the logic of the man who could not get new shoes on until he had worn them for some days. In fact when the IMPATT is placed in a resonant circuit

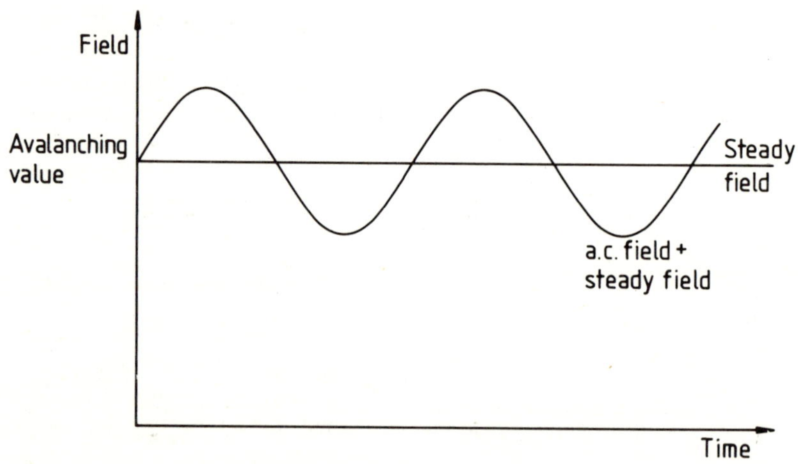

Fig. 7.6 First mechanism in IMPATT.

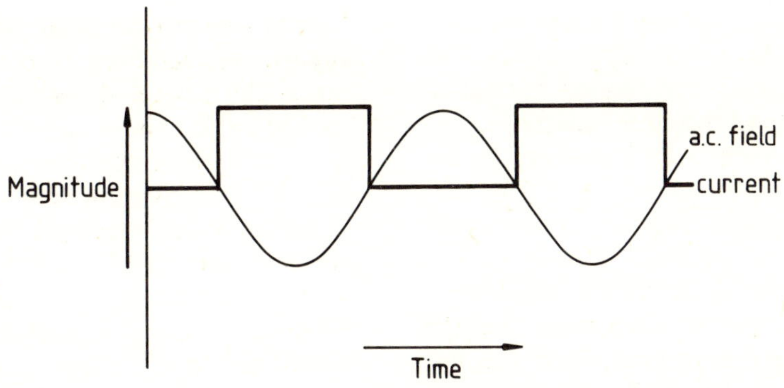

Fig. 7.7 Current and field in IMPATT.

it will generate its own a.c., the process being initiated by any small disturbance such as the pulse produced by applying the necessary working potential.

7.2.3 TRAPATT Avalanche Oscillator

The TRAPATT (Trapped Plasma Avalanche Triggered Transit) oscillator differs fundamentally from those microwave generators we have already considered. Basically these have all relied on anomalously conducting regions passing regularly through the device to give a cyclic variation in the overall resistance. In the TRAPATT the entire device switches regularly between a highly conducting and a highly resistive state.

If a high reverse bias voltage is applied to a p–n junction very rapidly, avalanche multiplication may set in, even though the applied voltage is less than the breakdown voltage. This results because a local high field region travels at a finite speed across the diode. (If the bias voltage were to be applied slowly the field would achieve its equilibrium value, which is less than that required for breakdown, at every instant of time (Fig. 7.8.) The high field region in its passage across the diode leaves behind it a dense 'plasma' of electrons and holes produced by the avalanching effect. The diode is

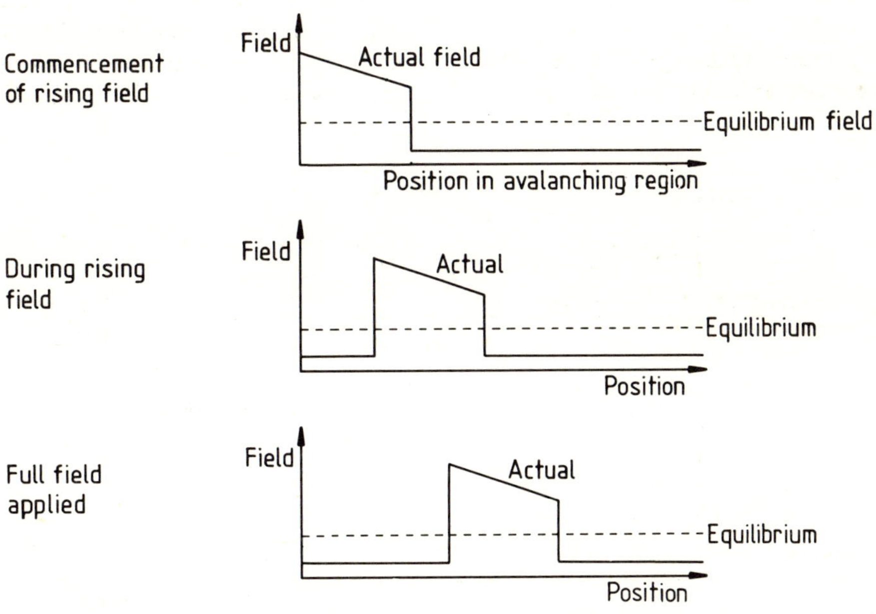

Fig. 7.8 Travelling high field region in TRAPATT.

thus in a highly conducting state and the current rises while, by the same cause, the voltage across it falls. The plasma then drifts out of the device restoring its original high resistance condition and the cycle restarts.

The frequency of this cycle is very sensitive to the external resonant circuit in which the TRAPATT operates and these circuits require sophisticated microwave engineering.

The essential requirement of TRAPATT operation is that the high field domain should travel faster than the maximum, or saturated, carrier velocity. Were this not so the carriers in the diode would be able to adjust to their equilibrium condition as the field moved.

Now when a rapidly changing field is applied to a semiconductor not only is there a flow of free charge carriers, giving the normal current, but the crystal structure changes slightly thus altering the positions of the 'fixed' ions. This constitutes a brief charge flow and is called a displacement current. The magnitude of this displacement current J_D is related to dE/dt, the rate of change of field with time, by

$$J_D = \varepsilon \frac{dE}{dt}$$

where ε is again the absolute permittivity of the material. If this current is greater than the saturation value of the normal current, the high field domain, which is carried by the displacement current, will travel faster than the free carriers, as we require. The condition then is that

$$\frac{J_D}{qn} > E_{sat}\,\mu$$

Therefore

$$\varepsilon \frac{dE}{dt} > E_{sat}\,\mu$$

and so

$$\frac{dE}{dt} > \frac{qnE_{sat}\,\mu}{\varepsilon}$$

Here n is the density of free carriers in the avalanching region before plasma carriers have been formed.

PROBLEMS

For values of required constants, see Problems to Chapter 1.

7.1 A typical Gunn diode is described in Section 7.2.1. If the ratio of the normal electron mobility to the heavy electron mobility is 5 and the critical domain field is 3×10^5 V m^{-1}, what is the length of the domain in this device?

7.2 What is the magnitude of the electric field outside the domain in the device described in Question 7.1?

CHAPTER 8

Noise

The final product of an electronic circuit is usually a signal which can be detected by either ear or eye. A trace on an oscilloscope, a picture on a television screen, music, speech or the position of a pointer on a meter are all various forms of required signal.

The processes which provide the final signal are each slightly distorted by the random and uncertain behaviour of electrons in the circuit components. The number of electrons available for conduction changes constantly as they become free, live for a short while and then collapse back to be immobilised on a fixed atom. Even during their free lifetime their path will be disturbed as they collide from time to time with one another and with the vibrating atoms of the solid in which they are travelling. Overall the effect will be to make the signal vary unpredictably about some average magnitude. This is manifest in the flicker of the meter needle, the lack of clarity in a television picture or the hum and crackle in the sound from a loudspeaker or earphone.

Because of this last, audible, effect the fluctuation superimposed on the average value of a signal is known as Noise, even when it is seen rather than heard.

Very sophisticated mathematical techniques are required for any considerable study of noise and this chapter deals only with some working principles.

8.1 FUNDAMENTAL ASPECTS OF NOISE

8.1.1 Introductory Definitions

The basic concept of noise as being undesirable sound has been extended, at least in technical literature, to mean any random variation in a quantity that is being measured or observed. The reason for this stems from the earliest technological, as opposed to scientific, experiences with the problem when it became manifest as a background hissing in the earphones of primitive wireless receivers. This real noise could obscure the wanted signal and eventually any disturbance of this nature such as the tremor on the needle of a sensitive meter or flashes on a television picture were designated 'noise'.

Sources of noise may be fundamental or man-made. Let us consider a very sensitive galvanometer. The suspension will be subject to incessant bombardment by air molecules and the vibrations this produces will set a limit to the smallest current that can be observed. This limit is fundamental. In addition, however, the bench on which the instrument stands will probably be subject to vibrations from external sources. This too sets a limit on the minimum detectable current but this is a man-made limit and can be changed by using sophisticated methods of suspension or damping. We will be concerned with fundamental noise here.

8.1.2 Origins of Semiconductor Noise

Noise in electrical circuits arises because all the mechanisms involved are in dynamic and not absolute equilibrium. Such processes are always varying with time although in such a way that over a prolonged period they show a fixed mean value. Some examples will make this clearer.

The free electrons in a bar of metal resemble gas atoms in a closed vessel in some respects. They are in incessant random motion at high thermal velocities and, since they carry an electrical charge, these random flights each constitute a small, local current flow. If the metal is not connected to a current-carrying circuit there can be no net flow of current through it and all the elementary currents must cancel out over a long period. To simplify the picture we can imagine the electrons rushing up to one end of the bar and then rushing back to the other end. If the bar is now made to carry a current from an external circuit this oscillation of

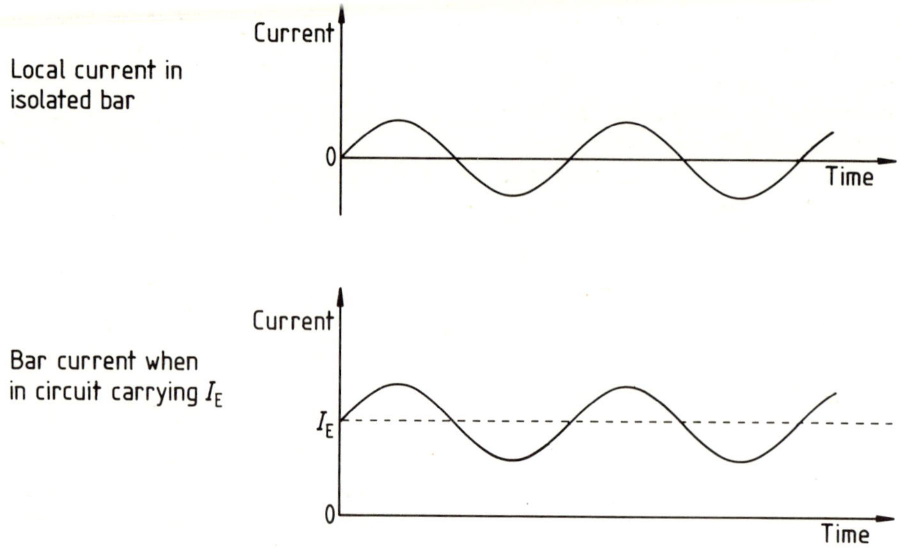

Fig. 8.1 Thermal noise.

carriers will be superimposed on the external current. Thus the required current has this oscillating noise added to it (Fig. 8.1). Because of its origins in the thermal motion of the carriers this is called THERMAL NOISE.

An irregularity that shows when electrons or holes pass through the depletion region of a diode junction stems from the fact that they will enter the region in a spasmodic, random manner. Suppose they were to be counted, starting at some arbitrary moment in time. At first the average current is zero and remains so until the first charge carrier comes along. If the counting is continued over a long time the step-like nature of the particle current becomes negligible and the averaged current approaches a steady value (Fig. 8.2). This phenomenon is known as SHOT NOISE likening the irregular arrival of individual carriers to that of lead shot on a surface.

Shot and thermal noise are both examples of what is known as WHITE NOISE. This name is given them because the noise current or voltage they yield is composed of equal contributions from currents or voltages varying sinusoidally with time at all frequencies and with random phases. The title 'white' refers to their similarity to white light carrying equal powers of all visible radiation frequencies. It is evident that there is no reason for any particular frequency to be favoured in these random mechanisms.

In contrast CURRENT NOISE, or $1/f$ noise as it is colloquially called, varies in power inversely with the frequency, f, which is being examined. Although there is no generally accepted origin for this noise it may arise from the capture and discharge of carriers from groups of traps, particularly those which exist in the interfaces between different materials used in a complete device. It is significantly absent from simple metallic resistors but in a transistor it makes a substantial contribution to the total noise spectrum, normally being the dominant factor below 10 kHz. One form of current noise which can exist only in a semiconductor or insulator is

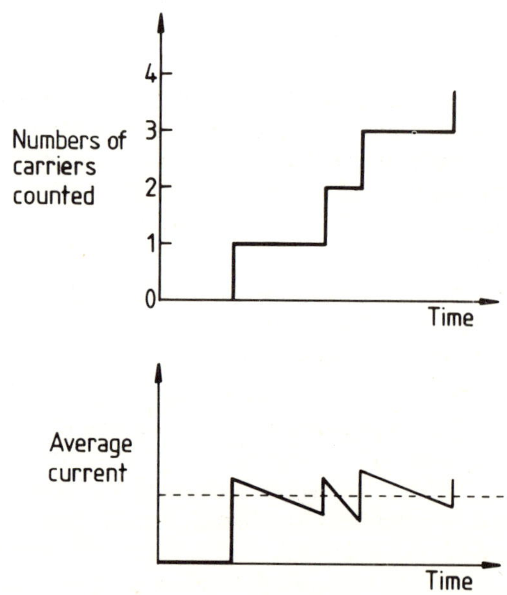

Fig. 8.2 Shot noise.

generation-recombination (GR) noise. The carrier population in a semiconductor is constantly varying as carrier pairs are born and recombine just as the population of a country varies constantly, between narrow limits, with birth and death. Since the conductivity of the semiconductor is proportional to its carrier density, this conductivity will be fluctuating and causing the current in the material to vary, thus producing the GR noise.

8.2 ANALYTICAL VALUES OF NOISE SOURCES

8.2.1 White Noise

If a resistance R is held at a temperature T, the available noise power W measured between frequencies f and $f + \mathrm{d}f$ is given by:

$$W = \frac{kT\,\mathrm{d}f}{\frac{kT}{hf}(e^{hf/kT} - 1)} \tag{8.1}$$

where k is Boltzmann's constant and h is Planck's constant. It is noted that this is independent of R but appears to vary with f. This latter prevents the noise being truly white but, except at high temperatures and low frequencies, the denominator is very nearly equal to one and, in general, it is regarded as negligibly different making Equation 8.1:

$$W = kT\,\mathrm{d}f \tag{8.2}$$

We need to determine what noise voltages and currents can be observed as a consequence of this noise power W. To do this let the 'noisy' resistor be represented by a noise generator of voltage V_n in series with the resistance R (Fig. 8.3a). A filter passing only frequencies between f and $f + \mathrm{d}f$ is placed in the circuit which is completed with a load resistance R_L. The current through R_L is $V_n/(R + R_L)$ and so the power W' dissipated in R_L is given by

$$W' = \left(\frac{V_n}{R + R_L}\right)^2 R_L$$

This has a maximum value when $R_L = R$

$$W = \frac{V_n^2}{4R}$$

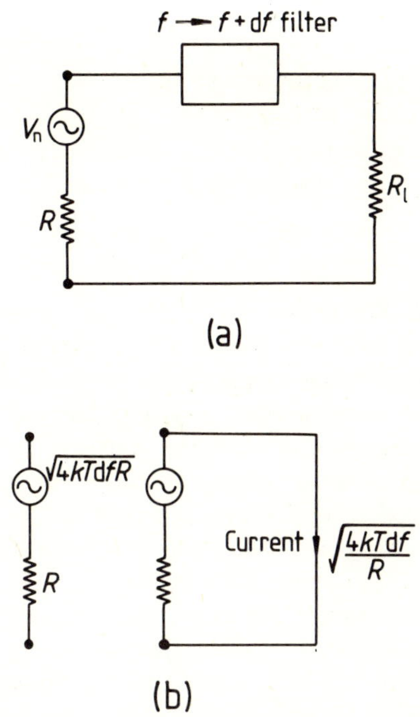

Fig. 8.3 Thermal noise.

Now the available power is $kT\,\mathrm{d}f$ and so

$$\frac{V_n^2}{4R} = kT\,\mathrm{d}f$$

or

$$V_n = \sqrt{(4kT\,\mathrm{d}f R)} \tag{8.3}$$

We can now represent the resistor by a source of voltage equal to $\sqrt{(4kT\,\mathrm{d}f R)}$ in series with a resistance R. If this equivalent is now short-circuited a noise current i_n will flow where

$$i_n = \frac{\sqrt{(4kT\,\mathrm{d}f R)}}{R} = \sqrt{(4kT\,\mathrm{d}f/R)} \tag{8.4}$$

(Fig. 8.3b).

The currents and voltages involved are very small as may be seen if we take a 100 ohm resistor at 300 °K and observe signals between 1 and 1.1 kHz.

$$i_n = 2\sqrt{\left(\frac{kT\,\mathrm{d}f}{R}\right)}$$

$$= 2\sqrt{\left(\frac{10^{-23} \times 300 \times 10^2}{100}\right)}\text{ A}$$

$$= 11 \times 10^{-11}\text{ A}$$

and

$$V_n = 11 \times 10^{-9}\ \mathrm{V}$$

The shot noise in a junction passing a current I may be expressed as a noise current i_{sn} where

$$i_{sn}{}^2 = 2qI\, df \qquad (8.5)$$

Equation 8.5 shows the shot-noise power to be proportional to the magnitude of the charge q on each carrier. The physical source of this proportionality can be seen if it were assumed, say, that the current were due to a few heavily charged particles when their random arrival would cause violent fluctuations in the time-averaged current shown in Fig. 8.2.

Before moving on to nonwhite noise we must see how thermal and shot noise can be related to a diode. When no impressed current passes through the diode, that is when short-circuited, we have two opposing currents, each equal to the saturation current, flowing in the junction. This should give a shot noise current of $\sqrt{(4qI_{sat}\, df)}$. However the noise current from a component of resistance R, with no impressed current flowing through, should be $\sqrt{(4kT\, df/R)}$. These two expressions will be equal if

$$R = \frac{kT}{qI_{sat}}$$

Equation 4.12 shows this to be so at zero bias. Both ways of computing the zero-bias white noise in the diode therefore give the same answer.

8.2.2 Frequency-Dependent Noise

Frequency-dependent noise, although of practical importance, may be regarded as a less universal problem than white noise. A metal resistance with good contacts would show no such noise since there would be no carrier traps either in the bulk or at the interfaces, nor would there be generation-recombination noise since the free-electron population has a constant value.

The $1/f$ noise in a semiconductor may be expressed as a noise current i_τ related to the external current I by

$$i_\tau{}^2 = K\frac{I^2\, df}{f} \qquad (8.6)$$

The dimensionless constant K must be determined by experiment and it is found to vary strongly with the surface treatment of devices.

Table 8.1 Noise Magnitudes and Sources

Name of noise	Magnitude of short circuit noise current	Source
Thermal (Johnson or Nyquist)	$\left(\frac{kT\, df}{R}\right)^{1/2}$	Thermal motion of carriers
Shot	$(2qI\, df)^{1/2}$	Irregular passage of carriers through circuit
$1/f$	$\left(\frac{KI^2\, df}{f}\right)^{1/2}$	Obscure—Possibly discharge and recapture in traps at interfaces
Generation-recombination	$\left(\frac{2\tau I^2\, df}{N(1+4\pi^2 f^2\tau^2)}\right)^{1/2}$	Fluctuations in number of free carriers

Generation-recombination (GR) noise current i_{GR} has an interesting form:

$$i_{GR}{}^2 = \frac{2\tau I^2\, df}{N(1+4\pi^2 f^2 \tau^2)} \qquad (8.7)$$

Equation 8.7 would apply to an intrinsic semiconductor sample with a carrier life time τ and a total of N carriers. (Note that N is the total, and not the density, of carriers in the sample.) This noise current is feebly dependent on the frequency when this is much less than $1/2\pi\tau$, but approaches a $1/f$ dependence at higher frequencies.

If the semiconductor is not intrinsic but obtains some of its carrier population from impurity centres then a number of carrier lifetimes will be operative. The form of Equation 8.7 will become more complex in this case.

Magnitudes and sources of the various noise currents are collected in Table 8.1.

8.3 NOISE IN PRACTICE

8.3.1 Noise in a Complex Circuit

As we have seen, all passive circuit components will generate thermal noise. In general there will be other noise sources in the component as well. These noise generators must be compounded by adding the separate powers and then extracting a total noise current or voltage as required.

If two components of resistance R_1 and R_2 and noise powers W_1 and W_2 respectively are connected in parallel (Fig. 8.4), the squares of the noise currents W_1/R_1 and

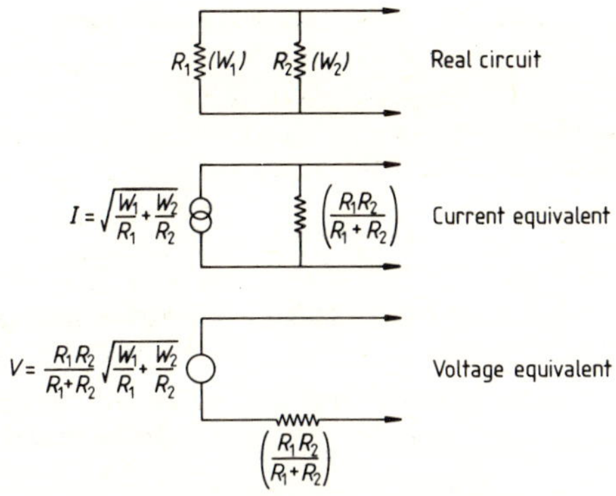

Fig. 8.4 Parallel noise sources.

W_2/R_2 may be added giving an effective current generator of magnitude

$$\sqrt{\left(\frac{W_1}{R_1} + \frac{W_2}{R_2}\right)}$$

and resistance

$$\frac{R_1 R_2}{R_1 + R_2}$$

If it is more convenient for the circuit analysis this may be replaced by a voltage generator of magnitude

$$\left(\frac{R_1 R_2}{R_1 + R_2}\right)\sqrt{\left(\frac{W_1}{R_1} + \frac{W_2}{R_2}\right)}$$

If the components are connected in series (Fig. 8.5), the noise voltages must first be added giving an effective voltage generator $\sqrt{(R_1 W_1 + R_2 W_2)}$ and resistance $(R_1 + R_2)$. The current generator equivalent to this voltage generator circuit is then

$$\frac{\sqrt{(R_1 W_1 + R_2 W_2)}}{(R_1 + R_2)}$$

and resistance $(R_1 + R_2)$.

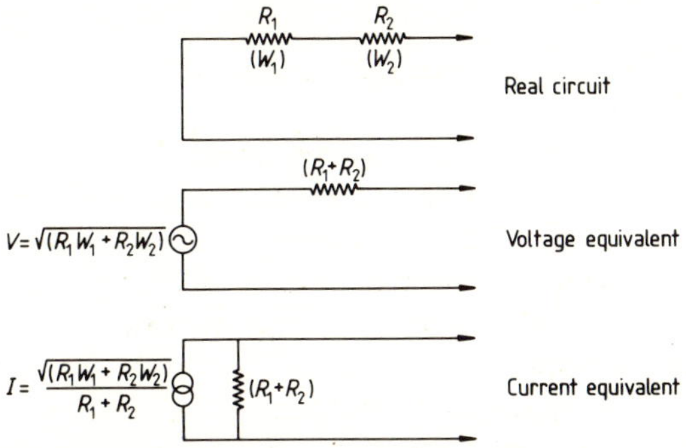

Fig. 8.5 Series noise sources.

By carefully applying these techniques any circuit may be finally reduced to a single resistance and voltage generator or resistance and current generator.

8.3.2 Minimising the Effects of Noise

Noise is a problem inasmuch as it tends to obscure the wanted signal in any indicator such as oscilloscope trace, meter needle, or loudspeaker. The important factor is the ratio of signal amplitude to noise amplitude, the S/N ratio, and it is this factor which must be maximised.

Signal amplification is not useful since any amplifier will add its own noise to that already in the signal. Suppose a signal S with noise N is amplified by a factor μ with the amplifier adding noise n then:

$$\text{Signal/noise ratio before amplification} = S/N$$

$$\text{after amplification} = \frac{\mu S}{\mu N + n}$$

which is smaller. We must then render the noise sources as small as possible. The inescapable noise is, as we have seen, thermal noise and it can be diminished by reducing the temperature of the component in question. Although the use of liquid nitrogen with its boiling point of 77 °K is quite widespread, substantial refrigeration is not always convenient, for many devices will suffer considerable, and sometimes irreversible, changes in their electrical characteristics with such drastic cooling. Both GR and $1/f$ noise are improved by current reduction and in addition these will improve if the working frequency f is increased. The last technique will eventually be limited by the failure of the component, or other parts of the circuit, to respond quickly enough for the high frequencies involved. All noises are diminished by reducing the bandpass df and a rather more interesting limitation obtains here, for any real signal will itself have a finite bandwidth, that is it will not be a perfect single frequency sine wave. Consequently as the value of df is reduced to remove noise it will, finally, start to reduce as well and the S/N ratio will not thereafter be improved.

Having reduced all the individual noise sources as far as practicable there remains a technique which will enable a signal to be extracted from a noisy background provided that the signal is cyclic or repetitive in form. This is known as BOX CAR INTEGRATION. It can be

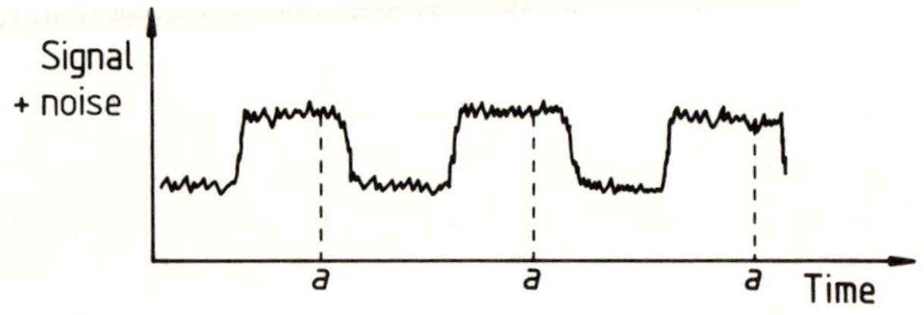

Fig. 8.6 Principle of box car integration for noisy square wave.

exemplified by considering a train of noisy, square pulses as shown in Fig. 8.6. If the height of the pulse is measured at some specific point *a*, it may vary widely from pulse to pulse due to the varying noise added to it. By taking and averaging a very large number of height measurements at points *a*, an increasingly accurate estimation will be obtained. Measurement and averaging is eventually carried out along the entire pulse and the average values displayed to give an almost noise-free picture of the signal shape.

PROBLEMS

For values of required constants, see Problems to Chapter 1.

8.1 The resistor in the worked example in Section 8.2.1 is cooled until the thermal current noise is 5.57×10^{-11} A. What is the temperature of the resistor?

8.2 What is the thermal voltage noise of the resistor in Question 8.1, measured under the same conditions?

8.3 If a diode is forward biased until it passes half the saturation current, by what factor will the shot noise be increased over that shown by the unbiased diode?

8.4 Four 10-Ω resistors are connected in series to form a continuous loop, what is the thermal current noise measured across any one of the resistors? Take the temperature as 300 °K and the frequency bandpass as 10 Hz.

8.5 The signal-to-noise at the input of an amplifier is 100. After an amplification of 10 the signal-to-noise ratio of the amplifier output has fallen to 95. What is the noise in the amplifier compared with the original noise in the signal?

Answers to Problems

1.1. 2.12
1.2. 0.0133

2.1. 7.29×10^{-10} m
2.2. 9.75
2.3. To within 0.146 m s^{-1}
2.4. 4.54×10^{24}
2.5. 1.98×10^{8} m^{-3}

3.1. 0.135
3.2. 10^{-3} s
3.3. 229 °K
3.4. 1.15×10^{6} V m^{-1}
3.5. 2.93×10^{6} m s^{-1}
3.6. 2.69×10^{-11} m^2
3.7. 10^{-10} s
3.8. 8.85×10^{-12} F m^{-1}

4.1. 10^{18} m^{-3}
4.2. 3.36×10^{-8} A
4.3. 7.46×10^{5} Ω
4.4. 2.40×10^{-12} F
4.5. 1.30 V
4.6. 2.25×10^{-12} F; 1.10 V
4.8. 88.2%
4.9. 3×10^{-7} m

5.1. 10^{-3} s
5.2. 2.06×10^{-12} A
5.3. 3×10^{-4} W
5.4. 0.1 eV
5.5. 1 eV
5.6. 12° 40′
5.7. 12° 20′

6.1. $$\frac{\rho a}{h\left\{t - \sqrt{\left[\frac{2\varepsilon}{Ndq}(V_0 + V_g)\right]}\right\}}$$

6.2. $$\frac{at}{\varepsilon_1 \mu h(V_{SD} - V_g)}$$

7.1. 1.67×10^{-6} m
7.2. 6×10^{4} V m^{-1}

8.1. 77 °K
8.2. 5.57×10^{-9} V
8.3. 1.12
8.4. 1.49×10^{-10} A
8.5. 0.526

Index

Risk

All firms exposed to it

For international business mgrs

it is topic of particular concern

Risk determined by a ~~number of~~ 3 main factors

Economic, -

Political, -

and legal factors -

Economic risk

is the...

likelihood that changes in econ. circumstances in the host country will affect the performance and return in that country from the point of view of the MNE

Political risk

is the...

Probability that changes in the politic climate of the host count will negatively effect a MNE profit ~~to that country to~~ or impede the attainment of other important objectives in that country

~~[illegible]~~

Factors affecting Economic risk

1. Changes in monetary and fiscal policies
2. The extent to which the host country can pay
3. off its external debt
3. rate of inflation

PRIFYSGOL CYMRU

UNIVERSITY OF WALES

AVAILABILITY OF THESES

The University regulations lay down that research work accepted for a higher degree shall be openly available, and subject to no security classification or restriction of access. Candidates are required to complete the attached form, signifying their consent for their thesis (if accepted) to be available for photocopying and inter-library loan, and for the title and summary to be made available to outside organisations.

It should be noted that the University requires the reader of a thesis to sign a declaration that in any statement or written work being prepared on the subject, or cognate subject, of the thesis due acknowledgement to the author will be made for any helpful information obtained from the thesis.

The attached self-adhesive form should be inserted inside the front cover of each copy of the thesis.
